SIX IMPOSSIBLE THINGS BEFORE BREAKFAST

LEWIS WOLPERT

Six Impossible Things Before Breakfast

The Evolutionary Origins of Belief

faber and faber

First published in 2006
by Faber and Faber Limited
3 Queen Square London WC1N 3AU

Typeset by Faber and Faber Limited
Printed in England by Mackays of Chatham plc, Chatham, Kent

A CIP record for this book
is available from the British Library

ISBN 978-0-571-20920-0
ISBN 0-571-20920-3

2 4 6 8 10 9 7 5 3 1

Acknowledgements

Grateful thanks to a number of people. Jack Herberg and Judith Landry commented on an early draft and Kim Sterelny, Philosophy Department, Victoria University of Wellington, New Zealand, made most helpful suggestions. Maureen Maloney typed the book – thanks again. And Alison Hawkes corrected and commented on draft after draft – invaluable help. Thanks too, to my agent Anne Engel and my editor Julian Loose needs many thanks, not least for the title.

Contents

Introduction

I became interested in belief for several reasons. In the first instance I wanted to know why my non-science friends had such difficulty with understanding science and why there was a quite strong anti-science movement. It was a real puzzle for me as I believe science to be the best way to understand how the world works. This led me to examine the unique origins of science among the Greeks, and its unnatural nature. Related to the general absence of belief in the scientific method was the belief in what I regard as the unbelievable, from angels to aliens to levitation and telepathy. How could people believe in things for which there seemed to be no reliable evidence? It really irritated me. And then there was religion, which affected me personally.

I was quite a religious child, saying my prayers each night and asking God for help on various occasions. It did not seem to help and I gave it all up around sixteen and have been an atheist ever since. Then my youngest son, who had been through a difficult late adolescence, was evangelised and joined the London Church of Christ. This is a fundamentalist Christian church, which takes the Bible literally. Contrary to what friends thought, I was not upset, as the church really helped Matthew. But the following incident reflects our relationship. Sitting in my office, Matthew said he was so envious of me, as I was so fortunate. Unused to receiving such a positive remark from any of my children, I beamed, and asked what he so envied. The reply was 'You are going to die soon, certainly before me.' I was shocked. Why was this so desirable? It was because he wanted to die so that he could go, he strongly believed, to heaven. In discussion, his position was totally rational and he could not, according to the religious rules, take his own life. I had to accept his position, albeit reluctantly. I did relate the incident to his sister Jessica. A week later I found a note on my chair 'Jessica says you think you are going to heaven when you die. We need to talk!'

My aim in this book is to try and understand what determines what people believe about causal events, and so it is on causal rather than ethical and moral beliefs that I will focus. A key issue is to determine what distinguishes our thinking and beliefs from other animals' and how this might have evolved. There will thus be a quite strong biological emphasis on how our brains function, together with an evolutionary viewpoint.

My aim is not to disparage the beliefs of others, even though I do not share them. This aim may not always be successful as I am neither religious nor do I have any beliefs in a spiritual world or paranormal happenings. My thinking is based on a belief in the scientific process, and the necessity for evidence. It is thus essential that from the very beginning I set out my own beliefs, even though I will try not to alter my material to fit in with them, or to try and persuade the reader to share them with me. I admit I am a reductionist materialist atheist.

I am committed to science and believe it to be the best way to understand the world. I am an atheist reductionist materialist. I know of no good evidence for the existence of God. I am in no way hostile to religion provided it does not interfere in the lives of other people or come into conflict with science. I do not believe in paranormal phenomena such as communication with the dead, telepathy, mind reading, ghosts, spirits, psi, psychokinesis, levitation – the evidence is just not there. I am thus similarly very suspicious about claims for the success of alternative medicine from reflexology to homeopathy and spiritual healing, and particularly psychoanalysis. While I am not concerned here to criticise or question beliefs in these topics, inevitably my views will come through.

A tool that I try to use is evolutionary biology, as this is a topic I am familiar with, though mainly in relation to the development of embryos. It is mainly evolutionary psychology that is relevant here, and I am aware that this is quite a controversial field. There are those who do not wish to believe just how much our genes determine our behaviour. They should perhaps reflect on lower animals, such as flies, who can land safely on the edge of a glass without practice, and birds who build wonderful nests. It is our genes that make the embryos from which we develop and end up as humans, and they determine how our brains will work. Yes, culture is important, as is nurture, but they both interact in and on a very complex biological system.

The key evolutionary idea related to our minds is that of adaptiveness; that is those behaviours, thoughts and beliefs that help us humans to survive better. Genes can determine variants in such processes and evolution will select those individuals that survive best, and will so select those genes. The problem is to identify just what those characteristics are and how genes affect them, and to distinguish them from those that arise from interaction with the environment and learning. Alas, much of the evolutionary biology that I will use is similar to Kipling's 'Just So' stories, like how the camel got its hump. It is very difficult to get reliable evidence to show whether one is right or wrong. One cannot go back in time, but I hope that this book, like Kipling's, is both interesting and entertaining.

I have only a limited illusion that I have provided useful insights into the nature of belief, but hope to have raised some important, if controversial, issues. I would be surprised, and disappointed, if some of the ideas did not provoke quite a few vigorous rejections, and alternative explanations.

And as for the title, it comes from Lewis Carroll's *Through the Looking Glass*, when Alice encounters the White Queen and they talk about belief. When Alice says she cannot believe in impossible things, the Queen replies: 'I dare say you haven't had much practice. When I was your age, I always did it for half-an-hour a day. Why, sometimes I've believed as many as six impossible things before breakfast.'

SIX IMPOSSIBLE THINGS
BEFORE BREAKFAST

CHAPTER ONE

Everyday

Believing passionately in the palpably not true . . .
is the chief occupation of mankind.
H. L. Mencken

Our lives are full of causal beliefs about events that do not fit
with our expectations – why the children are late from school,
why the car will not start, why the weather is so bad, and why
we have got ill. We humans have a basic need to have beliefs
that account for important events in our lives, and these can be
quite sensible and rational. We all have beliefs about how the
day-to-day world works, and it is some of these common and
quite simple beliefs, like those related to risks of various kinds,
that I want to explore first, particularly the strange causes that
we believe in, and why these beliefs are so persistent. The focus
here is largely on Western culture. I am not concerned at this
stage with more complex beliefs like those relating to religion,
the paranormal or health, which will be considered later, but
many similar principles are involved, and the boundaries may
be rather fuzzy.

There is a strong motive for explaining any phenomena that
affect us in causal terms, an ingrained need to organise the
world cognitively – both the external world and the internal
world of the individual. This cognitive imperative, which has
been called a belief engine, may have evolved because it was
essential for human survival, and an enormous aid to activities

such as finding food, making tools or avoiding danger, and so became instinctive. In one study, residents in an area where an earthquake had occurred during the night were asked 'What was the first thing you did when you felt the earthquake?' Almost all responded first by saying that they had wondered what had occurred and why, before talking about what they then did. This belief engine has served us well, and as we shall see, gave us technology.

Clifford Geertz, the anthropologist, points out that insufficient attention has been given to just what common sense is. He draws a distinction between our apprehension of reality and down-to-earth everyday wisdom. When we refer to common sense, we suggest that it is a matter of judging the world sensibly and effectively, and so coping with the problems of everyday life. If someone lacks common sense, it does not mean that they do not grasp that rain wets, or fire burns, but that they have not taken sensible precautions to avoid getting wet or being burned. For the Zande in Africa, as we shall see, it is when common-sense explanations fail that witchcraft is invoked.

A frequent feature of beliefs is that when examining evidence relevant to a given belief, people are inclined to see what they expect to see and conclude what they expect to conclude. We only become critical of information when it is clearly not consistent with our beliefs, and even then may not give up that belief. Moreover, confirmatory information or events are much better remembered and recalled than those that contradict what we hold to be true. It is, as Francis Bacon put it, that 'Man prefers to believe what he prefers to be true.'

One psychologist has suggested that the reason why we are so attached to our beliefs is because they are like our possessions. Like material possessions, they can make us feel good. Even the way we talk about beliefs is like the way we talk about things we own. We 'hold', 'acquire', 'inherit', 'give up' beliefs. But our beliefs are much more to us than possessions: they are part of our very identity. Criticism of our beliefs can feel like a criticism of ourselves.

Many beliefs emphasise goal and purpose in different contexts – key words are 'design', 'purpose', and 'made for'. It is thus quite hard for some people to believe that biological evolution does not have a purpose, and that evolution is based on chance events leading to variation followed by selection. Even the AIDS epidemic was believed by some to be a global punishment. There are a large number of what have been termed urban myths or beliefs. For example, some believe that when the hijacked plane crashed into the Pentagon only one thing survived – a copy of the Bible, and another example is the belief that all Jews were absent from work in the Twin Towers on September 11th.

In relation to causal events, the 'Why?' class of question is usually only raised when the event is abnormal or out of the ordinary. The death of a young Spanish matador called Yiyo when he substituted for another matador was much more upsetting to the public than if he had been killed in the normal run of his work. These feelings are typical of events that are preceded by an exceptional one, and so have a much greater effect. And in this case there was also the belief that it is bad luck to substitute for another matador. How have such beliefs about luck, particularly bad luck, arisen? One possibility is that they reflect recalling other 'if only' events, which are more readily remembered. How many of us believe that switching queues results in the one we left speeding up?

There is also anticipatory regret. Tests found that people would be less willing to sell their lottery ticket the shorter the time before the result was to be announced. There is also evidence that some people believe in the probability of an event by the vividness with which they can imagine it. Or again, take the hesitation to throw out old things on the basis that one is then bound to have an urgent need of them.

How much does logic influence the creation of beliefs? Deductive reasoning is the process of drawing valid conclusions from a particular set of premises – Euclidean geometry is the great example. But in daily life, beliefs about the conclusion can influence the validity of this process. If the logical conclusion is consistent with a person's beliefs about the world it makes the logical deduction easier. So subjects believed, correctly, the following to be valid nearly 100 per cent of the time:

> No cigarettes are inexpensive.
> Some addictive things are inexpensive.
> Therefore, some addictive things are not cigarettes.

By contrast only around 50 per cent thought the following is valid, even though it is logically correct:

> No addictive things are inexpensive.
> Some cigarettes are inexpensive.
> Therefore, some cigarettes are not addictive.

The suggested basis for this difference in getting the logic right is that it involves two different processes, and different regions of the brain. One is the formal rule-based process using the bilateral parietal system of the brain, while the other is a complicated automatic process based on left frontal and temporal lobes. Belief-laden and belief-neutral arguments can result from modulation between these two systems.

Just consider again:

> No unhealthy foods have cholesterol.
> Some healthy foods are fried foods.
> Therefore, no fried foods have cholesterol.

Valid or not? My guess is that your left frontal and temporal lobe belief system was activated, and so the conclusion seemed invalid. But it is valid, logically.

People's perceptions about themselves are peculiarly unreliable. The average person's belief about themselves is, in general, flattering. A large majority of the public believe that they are more intelligent and fair-minded, better describers and less prejudiced than the average person. This is as true of the general public as it is of university students and their professors. This set of beliefs is known as the 'Lake Wobegon' effect, after a fictional community in the stories by Garrison Keillor, where all the children are above average. Again, a survey of high school seniors in the USA found that 70 per cent thought they had above-average leadership qualities and only 2 per cent believed they were below the average. And of university professors, 94 per cent thought they were better at their jobs than their average colleague. Am I any different? Believe I am!

People also claim more responsibility for good deeds than for bad, and for successes than for failures. A study of young married Canadians found that they each overestimated how much they contributed to family well-being in terms of cleaning, childcare and so on. They also regarded themselves as less prejudiced than others. In a poll, 44 per cent of white Americans rated other whites as having more anti-black prejudice than they did, but only 14 per cent admitted to such prejudices themselves. People revise their own personal histories in ways that are influenced by their beliefs at the time that they recall earlier events. They both exaggerate the stability of their past beliefs – political ones, for example – and overestimate how much these have changed.

Even when quite reliable personality tests were used, subjects could not distinguish between their report about themselves and that of another person, particularly if it was flattering. There is research on the 'Pollyanna principle', which shows that there is a very common tendency for us to accept positive words in the report, rather than negative ones. Acceptance is higher if the individual is insecure and the tester of supposedly high status and expertise, even if an astrologer or graphologist. This emphasis on acceptance of the positive may be, for the individual, highly adaptive. If someone is told, erroneously, that they are better or worse at a particular task they will, in general, at once explain why. We are masters of the ad hoc explanation.

The so-called interviewer illusion provides another example. Many interviewers of people applying for jobs or places at a university feel confident in their ability to predict long-term performance. Research in this area shows that most interviewers overestimate their skills. I like a film company's verdict in 1928 on Fred Astaire's screen test: 'Can't act. Can't sing. Slightly bald. Can dance a little.' And a modelling agency's report on Marilyn Monroe – that she should learn to be a secretary or get married.

One of the interviewer's chief errors is the attribution of the person's behaviour to their inner disposition, their basic character, and the failure to take into account how they will behave in different situations. This is typical of how we judge the causes of other people's behaviour. There is a well-established tendency to account for our own behaviour in a rather different way from that which we use to explain the behaviour of other people. We believe our own behaviour to be heavily influenced by external causes and situations. By contrast, the behaviour of others is much more often believed to be the result of the product of their underlying personal traits and dispositions.

Personality tests provide an example of the Barnum effect: you can fool most of the people most of the time. People frequently accept as correct generalised, vague, and even bogus descriptions of themselves, if these descriptions are common to the population of which they are part. The core belief in astrology is that a person's personality is influenced by the state of the heavenly bodies at the moment of birth. The personality types resemble the namesake of the sign under which they were born – those under Capricorn, the goat, are stubborn and hard working, while those born under Leo, the lion, are proud and fearless leaders. All these associations are partly driven by simple representativeness-based thinking, which will be explained later. The same is true of graphology.

Making up a story to account for events in what seems to us to be a rational manner, is programmed into our brains, and is illustrated in split-brain patients, where the language ability in the left hemisphere is separated from the right hemisphere. A picture of snow in a field is presented by the left eye to the right hemisphere, and a bird's claw presented to the other eye. The patient is then asked to pick from a set of pictures those that are best related to what they have just seen. The left hand, controlled by the right brain, might choose a shovel for the snow and the right, controlled by the left brain, a picture of a bird. Then when the subject is asked why they made these choices, only the left hemisphere can control the verbal response. The chicken claw, says the subject, goes with the chicken and the shovel is needed to clean out the chicken shed. This is a made-up story, for the subject's left hemisphere cannot know about the snow – a nice example of confabulation, storytelling to account for the observations in a consistent manner but leaving out crucial bits of information. More about confabulation later.

Beliefs about the risks that affect our lives are very far from reliable. For example, in spite of numerous claims that seat belts save many thousands of lives every year, between 1970 and 1978, countries in which the wearing of seat belts was compulsory had on average about 5 per cent more road accident deaths than before the introduction of the law. In the United Kingdom, road deaths decreased steadily from about 7,000 a year in

1972 to just over 4,000 in 1989. There is no evidence in the trend for any effect of the seat belt law that was introduced in 1983; there's actually evidence that the number of cyclists and pedestrians killed increased by about 10 per cent. That twice as many children were killed in road accidents in 1922 as today, which many would regard as almost unbelievable, must not be taken as evidence that there is less risk when children play in the street today; rather, it almost certainly reflects the care taken by parents today in keeping children off the streets.

How are these beliefs about risks, which are both puzzling and shocking, to be explained? The answer seems to lie in our perception of risk and how we modify our behaviour accordingly. An important concept that has been developed by John Adams to account for people's handling of risk is the 'Thermostat Model'. An individual's propensity to take risks is influenced by their own experience and that of others, and this model assumes that the degree to which we take risks varies from one individual to another. The key feature in risk-taking is the balancing of perceptions of the risk and the possible rewards, and this balance may be a reflection of an individual's particular type of personality. In general, the more risks an individual takes, the greater will be both the positive and the negative rewards.

Of particular importance in the model is the level at which the thermostat is set. Those who are prepared to take risks have a high thermostat setting, while for others, who are more cautious, the setting is much lower. So, for example, a driver going round a bend in a road will be influenced by rewards and risks. These could include getting to an appointment on time, impressing his companion, his concern for his own safety and that of a child in the rear seat, the cost of damaging the car and of losing his licence, and so on. He will also have taken into account the condition of the road and the amount of traffic, as well as the kind of car he's driving.

A very important feature of this model is risk compensation – people modify their behaviour in response to what they believe are changes in risks to themselves. Thus, when we 'belt up', we may drive just a little more dangerously, so that the risk of an accident increases. And while the seat belt may increase our chances of survival, pity the cyclists, pedestrians and back-seat passengers. As Adams wickedly suggests, one way to prevent this, and make drivers drive with much greater care, is to fit all cars with a spike on the steering wheel, directed at the driver's heart.

There are at least three different kinds of risk. The first is easy to experience and recognise and can be called direct; an example is the danger presented by traffic or fears of being attacked. The second contains risks that have been identified with the aid of science; diseases like smallpox and cholera fall into this category. The third class contains what are known as virtual risks, those about which science is unsure, and includes BSE and global warming. Most of the studies that have been carried out deal with risks of the first two categories. In both of these, it is possible to calculate the risk associated with a particular activity like cycling in London or smoking. At least in these cases there is some objective information on which one can base one's behaviour.

Even where risk can be calculated objectively for any particular event, individuals' perceptions of that risk can be, and often are, quite different. For example, with respect to transport, much greater attention is given to accidents that occur in planes, ships or trains than to those on the road in cars, on bicycles, or when walking. There are two related reasons. In being transported by train or plane, you are putting your trust in the organisation or people who run them. You expect them to have taken all the necessary precautions. But when you drive a car, for example, which is on all grounds much more dangerous, you have the belief that you are in control. Thus the responsibility now rests on you or the driver, who is usually someone you trust. Yet the actual number of deaths for 1,000,000,000 kilometres travelled is less than one for airlines and trains, around five for car drivers and passengers, fifty for cyclists, seventy for pedestrians, and 100 for motorcyclists. These figures could be misleading as, for example, we walk far fewer kilometres than we fly, so that what might be perceived as a greater danger in walking is in fact not.

Social scientists have identified four different types of attitude that people have with respect to risk and how it should be handled: individualists try to control their own environment but oppose controls; egalitarians are also against controls, but regard nature as something to be obeyed; hierarchists believe that nature must be managed and are in favour of controls being imposed; fatalists just try to duck when necessary.

As far as sports are concerned, rock climbing is over 100 times more dangerous than skiing. Tennis, I'm pleased to say, has so few fatalities that there are no statistics. There is a 100 times greater chance of being killed by an accident in one's home than by a terrorist bomb. The likelihood of being killed by a fire in a public building is even smaller, but just look at the extensive fire regulations that must be followed in hotels and department stores, even though, compared to car travel, the risk of death from such fires is 4,000 times lower. Our perception of risk is complex, and may bear little relation to the objective probabilities. For example, surveys have shown that there is a perception among the general population that childhood vaccination is ten times more dangerous than it actually is, and that the possibility of death from a stroke is underestimated by a factor of ten.

Our common-sense approach to risk is clearly very unreliable. The risk of losing a few pounds on the lottery is very high, and most people know that; but it is hard to understand why they think that if the pot is greater, it is worth risking more. And who would not think it wise to bet on red in roulette if black had just come up ten times in a row? Some perceptions are quite irrational. For example, I know mothers who, when taking on a new au pair, are very frightened that she will run off with the child if there has been even one such case reported in the press. If I hear that a cyclist I know has been injured, I am much more cautious than usual, although my risk remains the same. Statistics have a much smaller impact on such risk assessments than they should.

A major error in personal risk assessment probably relates to our inability to make sound judgements when the amount of information is limited. For example, a little while ago there was a suggestion in the United States that a game called *Dungeons and Dragons* was risky since it might lead to teenage suicide. The evidence in support of this claim was that twenty-eight teenagers who regularly played the game had committed suicide. But the average suicide rate for teenagers nationwide in the United States is about one in 10,000. Since some three million teenagers played the game, the number of suicides that might be expected among the players was 300, so there was no significant statistical link between playing the game and suicide.

BSE presents a very clear example of the problems associated with virtual risks, where there is little hard objective data on which to base probabilities, and thus precautions. This leads people to behave even more freely, and in ways that reflect their personalities. For the idea that eating beef could be dangerous might, from a common-sense point of view, appear to be based on flimsy evidence. The key finding was the sudden appearance of a quite new form of Creutzfeldt–Jakob disease. It is generally accepted that BSE came from cattle eating infected sheep. The proposed infectious agent is believed to be a protein molecule called a prion, which is quite unlike a virus or bacterium. The anxiety about the future incidence of the illness comes from an understanding of the pattern of development of such diseases, which have long incubation times. Mathematical predictions of how many cases will appear over the next few years are complex, but could strike fear even into the heart of a sceptic. There may be just a few more cases, but it is just as possible that there will be thousands. No one knows.

BSE thus raises very difficult questions as to how such issues should be handled. Because of the possibilities of real dangers, the government has to balance the risks against all the social and financial costs of taking precautions. Fire regulations require enormously expensive constructions, though fires are rare. Building the Thames Barrier was a precaution for a relatively rare event. The social responsibility of scientists is to make public the information that can affect our lives, not to make ethical or political decisions as to what to do. It is precisely such difficult decisions that politicians must take, and it is also their moral and legal duty to take the necessary precautions.

Stopping at traffic lights during the BSE crisis, I mentioned to a fellow cyclist, a stranger, how dangerous cycling was. I had just narrowly escaped being run over. 'Yes,' he replied. 'But I bet you don't eat beef. Yet the risk of being damaged by cycling is so much greater.' 'Of course,' I said. 'But so are the benefits. I would be really depressed if I were to give up cycling, but giving up beef for a year or so until the evidence is clearer is no trouble at all. It's a matter of risk benefit.' Happily, he agreed, and we pedalled our separate ways, my helmet firmly on my head. Life is a risky business. But here again there may be some misplaced beliefs, as the real reason that those who wear helmets suffer so much less brain damage may be that they are, in general, more careful, and, for example, do not ride their bikes after drinking too much alcohol.

In general, people's beliefs and judgements are insufficiently sensitive to sample size. Estimating risk is a risky business, even for experts. Consider the problems doctors have to face when estimating the probability that one of their patients has a particular disease, even when the diagnostic test is positive. Here is the problem. The disease affects 1 per cent of the population and the probability of the test detecting the disease in someone who has it is 80 per cent. But the chance of a false positive – that is the test indicating that someone has the disease even though they do not have it – is 10 per cent. What should the doctor tell a patient who tests positive and asks what is the probability that they have the disease? Most doctors estimate the probability as around 75 per cent. Are they correct?

Back to our doctor: if we take a random 1,000 people, we should expect ten to have the disease. Of these ten, only eight will be detected by the test. Tests on the other 990 will give 99 false positives. Thus only 8 out of 107 with a positive test will have the disease. That is less than 8 per cent! The patient, contrary to what most doctors think, should be quite reassured.

Not all statistically relevant relations are causal, so how does one resolve this issue and distinguish between correlation and causation? The drop in the reading of the barometer correlates with, but does not cause, storms. One needs to distinguish genuine from spurious causes. One possibility is that the issue is resolved when one knows the underlying mechanism: a drop in atmospheric pressure causes storms and also causes the barometer reading to fall. This type of thinking is essentially based on mechanical principles and involves concepts of forces and power – basic scientific knowledge. But why, for example, do so many people believe that infertile couples who adopt a child are more likely then to conceive a child than infertile couples who do not adopt? They have some belief about the adoption causing fertility by unknown means.

Causal illusion has been demonstrated in a simple experiment. It is clear that when one billiard ball hits another it causes it to move, but when people are shown coloured discs moving on a screen they somehow cannot help believing that when one collides with another it pushes it away. Even dots on a screen can appear as chasing or avoiding each other. In a film in which one square approached another stationary one and the latter moved off without any contact having been made, adults had the impression of a causal effect. Adults' causal beliefs also follow the rules of contagion – once in contact always in contact – and the idea that the image equals the object. Thus, drinks that come in contact with a sterilised dead cockroach are avoided, and food that looks disgusting, like fudge made to resemble dog faeces, is rejected.

There is evidence that the very concept of 'self' can be very different: the Western view of a person as a bounded, autonomous and unique system can be quite alien to other cultures, where the influence of the surrounding society is seen as paramount. The considerable differences that exist between different cultures can affect their basic beliefs about the world, and even the way that they think. This is because social organisation directs attention to certain features and away from others, and can thus influence beliefs about causality. For example, in ancient Greece much emphasis was put on the individual, who could often choose the way they thought, and this was probably a major feature that enabled the Greeks, alone amongst all cultures, to discover scientific thinking. By contrast, the ancient Chinese felt that individuals were essentially parts of a close-knit community, and there were numerous obligations to family, friends and emperor. Debate was discouraged. Yet technologically, they were much more advanced than the Greeks; this technology was not based on science, but on imaginative trial and error. Their beliefs were holistic, quite unlike those of the analytical Greeks, as we shall see later.

Even if all cultures possess the same basic cognitive processes, the 'tools' of choice for the same problem may be quite different, just as a workman can use different tools for the same job. These special ways of thinking have persisted. Compared to Europe and the USA, China and other East Asian societies remain committed to the idea of the individual as less important than the society. There are studies that show that this can affect the allocation of attention, as Easterners see wholes, Westerners parts. Easterners see relationships amongst the parts but find it harder to concentrate on an object when it is embedded amongst others. Westerners' perceptions will be affected by their basic belief that they have control over their environment. When presented with scenes of underwater life, Americans focus on the fish, Japanese on the lake and the relationships between the fish.

The important implications of these and other studies is that people attribute causality to events they pay attention to. Thus Westerners should attribute causality to objects, Easterners to content and situations. It has been found that Americans interpret the behaviour of others in terms of individual traits like kindness and recklessness, whereas Hindu Indians explain comparable behaviour in terms of social roles like obligations and the physical environment. Americans thus perceive the cause of murder as mental instability, whereas Chinese see it as reflecting society's failure.

There are also studies showing that American students are more willing than Korean students to set aside beliefs based on experience in favour of logical explanations, and the latter are more likely to judge valid arguments as wrong if they have implausible conclusions. Both groups have similar abilities in logical thinking, but differences in response when logic conflicts with everyday beliefs. In addition, Westerners respond to conflict by trying to decide who is right, while Easterners tend to yield points to both sides. Easterners prefer holistic approaches and compromise, and show a willingness to accept contradictory arguments. The practice of using feng shui for choosing building sites, even for skyscrapers in Hong Kong, reflects the holist idea that causes and outcomes are very complex and interactive.

Illusory correlations are quite common, and illustrate a Chinese proverb: 'Two-thirds of what we see is behind our eyes.' When told of a cloud-seeding experiment that might result in rain, and told each day whether it rained and whether the clouds were seeded, those who believed in the effect were more likely to recall days on which both seeding and rain occurred, even when the information was random. Henry Thoreau said that 'We hear and apprehend only what we half already know.' Again, male students having telephone conversations with women they did not know, but believed to be attractive from photos given to them, resulted in the 'attractive' women having more friendly and warm conversations – the men's erroneous beliefs became self-fulfilling.

Many of us have superstitions. I do not myself like to proclaim that I am too well or happy for, irrational though it is, I do not like to tempt the gods. Others carry lucky charms or touch wood (I do it myself!). More than 200 years ago Gilbert White observed that 'It is the hardest thing in the world to shake off superstitious prejudices; they are sucked in as it were with our mothers' milk.' Their basic characteristic is that they are irrational beliefs founded on ignorance or fear and held with an obsessive reverence, particularly for omens and charms. Luck is also a key element, together with a belief in magic. The majority of superstitions are concerned with explaining or avoiding bad luck, rather than attaining good things.

Among the most common current superstitions are that it is unlucky to walk under a ladder, to break a mirror, to spill salt, or to open an umbrella indoors. In the USA, almost half the adults questioned 'touched wood' and avoided walking under a ladder. The mythology of fortune surrounding Friday the 13th is far from clear. The first reference in English only goes back to 1913. It is claimed to be related to biblical events such as the number of guests at the Last Supper and the crucifixion on a Friday. The fear of this day was shared by Napoleon, Churchill, and Franklin D. Roosevelt, who would actually cancel appointments on that day. Dice players talk to the dice, blow on them, and put them to their ears. My own father believed that he was continually being dealt bad hands at bridge, and others avoided being his partner. We all have a tendency to attribute inexplicable events to some sort of 'mystical' cause, like bad 'luck' or even, light-heartedly, gremlins.

Gremlins provide a nice example of the value of such beliefs. In the last world war, Royal Air Force crews ascribed mechanical failures to gremlins. It was half-jokey, but it did deflect blame from the maintenance staff, and, very important-ly, it promoted solidarity in blaming the gremlins for trying to defeat them. Consider, too, the beliefs sportsmen have about what they must or must not do before a game. There are all sorts of superstitions in baseball that affect batters rather than fielders. Actors say 'break a leg' before a show to wish their fel-low actors luck. Michael Crichton, author of 'Jurassic Park', makes a point of eating the same food for lunch every day when working on a novel.

A scientist once visited the offices of the great Nobel Prize-winning physicist Niels Bohr in Copenhagen, and was amazed to find a horseshoe was nailed to the wall over his desk, so he said to Bohr: 'Surely you don't believe that horseshoe will bring you good luck, do you, Professor Bohr?' Bohr replied: 'I believe no such thing, my good friend. Not at all. I am scarcely likely to believe in such foolish nonsense. However, I am told that a horseshoe will bring you good luck whether you believe in it or not! How can one argue with such logic?'

We have strong beliefs about certain events that we have experienced, and that were important to us, but how reliable are our memories? When a person experiences something extremely important or upsetting, the memory of the event is not recorded like a video: the process is much more complicated – just bits and pieces are taken in and stored. When later relating the experience, these are pieced together in a story. Indeed, false memories can quite easily be triggered by suggestion, and can be related to confabulation. This has important implications for those relying on the evidence of witnesses to violent crimes.

There are studies showing that children and teenagers can be induced to remember the experience of being lost in a shopping mall when young, even though it never happened. With time, these memories become more vivid and are rather like the 'repressed' memories of a trauma unearthed in psychotherapy. In fact, the techniques used by therapists for unearthing such memories of early abuse can also be used to create false memories.

In one experiment, Italian students who did not think demonic possession was possible were given articles to read that claimed it to be plausible and not uncommon. They were then asked to fill out a questionnaire about fear and anxiety and then some were told, falsely, that their particular set of fears indicated that they had probably witnessed demonic possession as a child. On a follow-up interview, nearly one fifth indicated that they had witnessed demonic possession as children. A number of the others also took a more positive view of the plausibility of demonic possession.

In another experiment, students from MIT were given the

biography of a guest lecturer. One group had a description of the speaker as cold, while to the other group the speaker was praised for his warmth. After the lecture, the students rated his performance and character largely in terms of the description they had been given beforehand.

It is also a quite general feature that we are overconfident in the correctness of our judgements. In a classic experiment by Solomon Asch, a group of six subjects was shown a line and asked which of three other lines was of the same length. Of the six subjects, five had been briefed to choose one of the wrong lines, and most gave their choices before the true subject made a choice. That subject then often chose the wrong line rather than the one that their own observation suggested was the correct one.

It is very hard to alter someone's beliefs, and we all look for confirmation rather than falsification. When students who were either for or against capital punishment were shown the results of recent studies, one confirming, the other challenging their existing beliefs, both groups readily accepted only that study that confirmed their existing beliefs. In another experiment, participants were initially told a story that showed that a risk-taker was a very good firefighter, while another firefighter who was much more cautious was mediocre at the job. Given this information, the group concluded that risk-takers were better firefighters. They clung to this view even when told that the stories were just made up. Again, a magician performed fake psychic phenomena before a group of students. Afterwards they were asked whether they believed he had psychic powers, and about two thirds of the students believed that he did. They were then told that he was just a good magician and that he was faking it and he had no psychic powers. But about half of the students still believed that he really did have them.

Once beliefs are formed, they tend to persevere and are very hard to change. Freud put it beautifully: 'To begin with it was only tentatively that I put forward the views that I had developed . . . but in the course of time they gained such a hold of me that I can no longer think in any other way.'

But why are our beliefs so hard to lose or change or give up? I think there could be an evolutionary explanation: if beliefs that saved lives were not held strongly, it would have been disadvantageous in early human evolution. It would be a severe disadvantage, for example, when hunting or making tools, to keep changing one's mind. Even more so when one was in danger, since it was necessary to pursue a course of action that would lead to safety. If one believed that unstable rocks were dangerous, it would have been wise to stick to that belief.

Many current beliefs do not have the same life-or-death context, and so we are for the most part able to indulge in all sorts of beliefs, and freely use the evidence we bring to support them, no matter where it comes from. This may account for the wide variety of causal beliefs that humans hold. These beliefs, I shall argue, had their origin in evolution in relation to tool use, but with cultural evolution they have evolved in most interesting and not easily understandable ways. The freedom to have beliefs is very important, but it carries with it the obligation to carefully examine the evidence for them.

All this will now be explored. What is the evolutionary origin of human causal beliefs, the origin of religious beliefs, beliefs about the paranormal and health, and finally, the special nature of scientific beliefs?

CHAPTER TWO

Belief

This act of mind has never yet been explain'd
by any philosopher.
David Hume 1739

The word belief, while freely and widely used to account, for
example, for causes in the previous chapter, is nevertheless not
easy to define. Neither philosophers nor scientists have been
successful. David Hume, my hero philospher, said of belief that
he regarded it as a great mystery. And some 200 years later,
Bertrand Russell recognised that belief was a central problem in
the analysis of mind. For many, belief is intimately associated
with religion, and religious beliefs will be dealt with in detail. It
is the everyday use of the word that I deal with in this book, and
I will focus on those beliefs that relate to the causes of events
that affect our lives in significant ways. Beliefs relating to moral
issues will receive much less attention.

The anthropologist Rodney Needham has analysed the origins
of the word belief. In Middle English, from around the twelfth to
the fifteenth century, the verb *bileven* already had the sense of
believing in a religion, of being valid or true, of having a conviction. Going further back, one finds *galifan* in Old English and
galaubjan in Gothic – *laub* is related to the Indo-European to
love, want, desire. The English concept of belief is clearly linked to
Christianity and the acceptance of the Christian faith.

Needham finds it so difficult to define reliably what we mean by belief that he seems almost tempted to abandon the word altogether. He does not do so, however, but points out that statements about belief need to be analysed within the framework of the different cultures, and a mastery of the local language, including its subtleties, is essential. For example, he says that the term *kwoth* for the spirit of the Nuer religion in Africa is very hard to understand. Needham goes on to say that 'Belief is not a discriminable experience, it does not constitute a natural resemblance among men, and it does not belong to the "common behaviour of mankind."' Thus, he argues, when one talks of the beliefs of other people, particularly in relation to non-Western religions, one has little idea of what is going on in their minds. I am not at all sure Needham is right, for all cultures have beliefs about causes, but whether there are similarities in different cultures as to their beliefs, and how they arise, is a central problem.

Another problem, as Needham points out, is how belief affects the behaviour of the individual. For example, a Saultaire Indian may endanger his life by believing a bear can understand what he is saying, and an Australian Aborigine believes that he will feel pain if a lock of his hair, taken away, is cut. How can such beliefs be reconciled with their experience? Belief is itself a word not always easily translated from one language to another.

The *Shorter Oxford Dictionary* gives two definitions of belief: 'The mental action, condition, or habit, of trusting to or confiding in a person or thing; trust, confidence, faith.' A second definition is 'Mental assent to or acceptance of a proposition, statement, or fact, as true, on the grounds of authority or evidence; the mental condition involved in this assent.' And for 'believe', 'To have confidence or faith in, and consequently rely upon.'

A key feature characterising something as a belief as distinct from factual knowledge is how reliable the evidence is for the belief. But while evidence is a key word in relation to the validity of causal beliefs, it too presents severe problems of definition. How does one obtain reliable evidence? By authority? By direct observation? By science?

A distinctive feature of belief is that it may be graded true or false to varying degrees, depending on the evidence available, but it usually carries conviction. Unlike common knowledge, beliefs always have a true and false value – how right or wrong they are – regardless of whether or not the individual is aware of this. Reliable knowledge, by contrast, refers to what is clearly known – how to drive a car, for example, or that this is a page in a book, are facts. Probably the same could be said of all, well nearly all, of mathematics. There cannot be anyone who could dispute the validity of Euclid's planar geometry unless they were to totally abandon rationality. And as we shall see, much of science can be considered to be reliable knowledge.

But belief sometimes comes close to knowledge for the individual – for those, for example, who have seen ghosts, a belief in their existence becomes knowledge, though to others it is unbelievable. Knowledge provides us with a disposition to behave in a way that is constantly subject to modification – we know where we are when we are walking home – while belief is a disposition to behave in a manner that may be quite resistant to correction by experience – it is quite safe to drive after a few drinks. Beliefs can be very strong, with little reference to knowledge or evidence, and it is often other people's beliefs that seem the most fallible.

Beliefs are held about factors that have an important effect on our lives: why we get ill, what will happen when we die, how someone we love will respond, how to change things in our environment to our advantage. Memory can be viewed as a type of belief, not least because it can be unreliable. Indeed memory itself can be shaped by current beliefs. For example, in relationships, individuals' recall of important events, especially those involving some sort of conflict, can be greatly at variance.

A common characteristic of beliefs is that they explain the cause of an event, or how something will occur in the future. It is causal beliefs that are the main focus of this book since they have a particularly strong influence on human behaviour. Many beliefs guide the way a person chooses to behave. Belief in God can make a difference to a person's behaviour – that is almost a definition of a true belief, for if it did not influence how the person behaved, it would have no consequences and would be irrelevant. As the French philosopher Descartes put it: 'I needed in order to determine what people really believed to notice what they did rather than what they said.' David Hume also emphasised how beliefs affected our actions: 'Neither man nor any other being ought ever be thought possest of any ability, unless it be exprest and put in action.' This fits very nicely with the view of the evolution of the brain in relation to action that I will propose.

In 1739, David Hume put forward his doctrine about causality. Our idea of causality, he claimed, is that there is a necessary connection among things, particularly actions. However, this connection cannot be directly observed, and can only be inferred from observing one event always following another. He thus argued that a causal relationship inferred from such observations could not in fact be rationally inferred. While this may be philosophically true, it is a problem for philosophers which need not concern us, as it is obvious what the cause is if, for example, I cut my hand with a knife.

David Premack, a psychologist, has pointed out that there are two classes of causal beliefs. One, as Hume suggested, is based on one event being linked to another, and can be called weak or 'arbitrary', for there need not be any obvious connection between them, like switching on a light. Animals can learn connections by the pairing of events through this process of associative learning. The other, which is uniquely human, is strong or 'natural' causality, and is programmed into our brains so that we have evolved the ability to have a concept of forces acting on objects. Such strong causal beliefs are already present in human infants, whose beliefs are described in the next chapter. A key question is how this type of belief evolved. Causal beliefs are a fundamental characteristic of humans; animals, by contrast, as we shall see, have very few causal beliefs.

Beliefs come from a variety of sources that include the individual's experiences, the influence of authority, and the interpretation of events. At their core, beliefs establish a cause and effect relationship between events, and thus can be used as a guide to how one should behave in particular situations and how to judge the behaviour of others. From an evolutionary point of view, beliefs should help the individual survive, and I will argue that they had their origin in tool making and use. But beliefs are much wider in their nature, and often serve to make the person feel better by, for example, promoting self-esteem, and providing satisfactory explanations for events that are not well understood.

In the following chapters I will distinguish between several kinds of belief in relation to the topics they deal with. The distinctions are not hard and fast, and the boundaries are often fuzzy. I will start with the acquisition of causal beliefs by children and then compare them to animals, and will argue that the latter rarely have causal beliefs. Then, considering tools, I will claim that the origin of human causal beliefs is related to tool use and manufacture. Once there were causal beliefs for tool use then our ancestors developed causal beliefs about all the key events in their lives, and I will examine the proposed mechanisms by which we now acquire our beliefs. False beliefs that result from abnormalities in the brain like confabulation and schizophrenia can provide insights into the way normal beliefs are formed. Religious beliefs that had their origin in attempts to account for crucial events in our ancestors' lives will be given special attention, together with the possibility of their being genetically programmed.

Then there are also paranormal beliefs like astrology and witchcraft, which vary across cultures, and for which evidence is rarely required. Beliefs about health, which are particularly important, also vary enormously, and are all too seldom based on proper evidence. Political and moral beliefs can determine how societies behave and include democracy, communism and racism. Scientific beliefs, which had their unique origins in Greece, have a special validity and are not personal but shared by the scientific community. A key question is: how similar or different are the belief processes involved. Finally, we may ask what sort of beliefs does the future hold?

I will be putting a big emphasis on the biological basis of belief, and also on evolutionary aspects of human behaviour. A very useful set of principles in relation to biological explanations was put forward by Niko Tinbergen, one of the founders of the scientific basis of behaviour in animals. He argued that behaviour can be explained in four separate but related ways: the physical cause, how it develops, its function, and its evolution. Consider, for example, hunger and sadness. The basic mechanisms, the physiological bases, are those that cause us to feel sad or hungry; how these develop – both in the embryo and after birth – is the second question; then there is their function and advantage to the individual: hunger to ensure eating, sadness to make up a loss; and finally, how did these behaviours evolve? It is such questions that we need to consider in relation to belief.

It may be helpful to briefly explain the role of genes in evolution, in order to clarify what I mean when I refer to a particular character, even a belief, being genetically determined. Genes are unique in that they are the only elements in the cell that replicate, and thus can be passed to successive generations. The genes provide a programme for the development of the embryo by controlling how the cells in the embryo behave to give rise to the adult. Genes are basically boring and passive, as they do nothing but provide the code for making proteins, which are the true wizards of the cells. But they provide a programme for where and when particular proteins are made and so control, for example, how the cells of the brain connect with each other, and how one region of the brain connects with other regions. Thus, in evolution, changes in genes can result in changes in the form and behaviour of an organism; and depending on whether or not it is adaptive, that is, leads to better survival, that change in the genes will persist. That genes can determine behaviour is evident when one looks at the enormous variety of behaviours that animals are programmed to carry out, from sex to nest building.

Evolutionary psychology is based on the idea that in evolution the human brain acquired a number of specialised computational mechanisms – sometimes called modules – that determine emotion, reasoning, pattern seeking, and so on, and thus affect what we believe. The modules may not be physically separated in the brain, and there is considerable fluidity. Michael Shermer, who has thought deeply about these issues, considers modules like these to underlie what he calls the belief engine. There may be a causal operator in the brain that compels us to try and find out why things that matter to us happen. Without this imperative, we would not be successful in developing technology, on which we are so dependent.

An inability to find causes for important events and situations leads to mental discomfort, even anxiety, so there is a strong tendency to make up a causal story to provide an explanation. Ignorance about important causes is intolerable. This also makes sense from an evolutionary perspective, as our ancestors needed to account for events rapidly even when they had little knowledge – delay could be a great disadvantage.

Belief has not, unfortunately, been the subject of much neuro-scientific research, and so the nature of the brain mechanisms that give rise to beliefs are poorly understood, though the study of mental illnesses that give rise to false beliefs may help. All this makes it difficult to know how the brain generates beliefs, as well as how to change people's beliefs – and as we all know, that can be very difficult.

One evolutionary approach to beliefs might make use of the meme, a concept introduced by Richard Dawkins in 1976. It refers to a unit of complex ideas, a unit of information that plays a role analogous to genes. Memes are memorable and can be likened to mind viruses. It is claimed that memes replicate and the selection of one meme over another may have no advantage to the individual in whose mind it rests. Since memes are claimed to have variation, heredity, and differential fitness, they could have the necessary properties for evolution by natural selection. Dawkins even claimed that we do not choose our memes, but that they choose us and manipulate us to their own ends. Just what a meme is, and how it is distin-guishable from beliefs, I find difficult. Is the word 'bird' a meme, and is the second law of thermodynamics also one? Apparently the song 'Happy Birthday to You' is a meme. There is no distinction made between memes relating to belief and knowledge. Moreover, no mechanism is proposed for the so-called replication of memes, or what they are selected for. Nevertheless, memes raise important questions on how social learning occurs, and why certain memes are so stable. Under-standing memes could help in understanding beliefs.

To understand the evolutionary origins of belief, it helps to understand what the brain is for. Belief is a property of the brain, which is made up of billions of nerve cells whose func-tion is totally dependent on the signals between them. But what is the primary function of the brain itself? I believe it has just one: to control bodily movements; and so this must be at the core of any attempt to understand belief. The evidence comes from the evolution of the brain.

Movement was present in our ancestral cells which gave rise to multicellular organisms some 3,000,000,000 years ago. They could move either by using flagella and cilia, whip-like structures that are a bit like oars, or by amoeboid movement, the cells extending processes at their advancing end, and then pulling themselves forward to where these attach. This movement was a great advantage in finding food, dispersal to new sites, and escape from predators. A key point is that the protein molecules that produced these movements are the precursors of all muscle cells. Muscle-like cells are found in all animals, including primitive ones like hydra, a small freshwater creature with just two layers of cells arranged in the form of a tube, which uses the movement of its tentacles to capture prey.

In higher forms, like flatworms and molluscs, muscles are well developed and the ability to move is a characteristic of almost all animals. One only has to think of such forms as diverse as earthworms and squirrels. Again, this ability to move is fundamental to animal life – not just finding food and shelter, but the ability to escape from enemies. And this is where brains come from. The first evidence for brain-like precursors is the collection of nerves that are involved in controlling movement, like the crawling of earthworms or flatworms. Getting the muscles to contract in the right order was a very major evolutionary advance, and required the evolution of nerves themselves. Here we find the circuits of nerves that excite muscles in the right order: the precursors of brains.

The first advantage of the ability to move was most likely dispersal and finding new habitats, but once the ability to move had evolved, it opened up new advantages such as finding food and avoiding danger. It became necessary to perceive the nature of the environment in order to decide when and where to move. There was a need for reliable senses. Light-sensitive cells are present among single-cell organisms so it is not too difficult to imagine light coming to control movement. Then, later, came the eye. Of course there were other sensory systems that could detect touch, temperature and odours. All these had and have but one function, to provide information for the control of movement. Emotions evolved to help animals make the appropriate motor movements like flight, attack, or sex. And that is why plants do not have brains. They are very successful but they do not need brains for they neither move significantly, or more importantly, exert forces on their environment in order to modify it for their own survival. No muscles, no brain.

There is no human or animal emotion that is not ultimately expressed as movement; in fact the argument is somewhat circular, for what else is human behaviour? Sense organs have only one function, to help the organism decide how to move. Once the brain developed, it took on other functions such as those related to homeostasis, like hormonal release and temperature regulation. Our brain, with all its imagery and memory, somehow enables us to decide how to behave, and it has only one ultimate function and that is to control bodily movements. The evolution of the brain that gave us beliefs is no more than an expansion of the original circuits that controlled movement in our ancient animal ancestors.

An internal representation of self arose in evolution from coordinating inner-body signals to produce appropriate behaviour. Increased accuracy and planning of movements was achieved by having mental models of the body in relation to the environment, and realising how these were causally related. Thus, when a stone falls on one's toe, one knows it, and one has to decide how to respond. More generally, as David Hume made clear, there is no experience of 'self' as something distinct from our body. Given that the main function of the brain is to control movement and to choose the appropriate movements for survival, it is not that unreasonable to suggest that belief arose in relation to tool use and manufacture, as both require a belief in causal interactions. This is a different view from that widely held, namely that the evolution of the human brain is related to social interactions. It will be helpful to look first at how human children acquire their causal beliefs, and then to compare them with those of animals.

Children

Children

Causation is the cement of society.
David Hume

To what extent are children's beliefs about cause and effect innate, and how much is due to their experience of the world? By innate, I mean that the human brain develops in a particular way so that it will readily acquire causal beliefs about how the physical world behaves, unlike the brains of other animals. This in turn means that brain function in relation to belief has a strong genetic basis. Genes and the environment interact, of course, and human infants and children have many of the same early experiences as primates in terms of their own early movements and what they see. But they interpret them so very differently. Causal understanding in children is what developmental psychologists call a developmental primitive – it develops early in life, is initially little related to culture although greatly influenced by interactions with others, and is achieved with a minimum of effort. While many psychologists avoid calling it innate, it clearly is. From an early age children begin to understand causes that affect objects, and those that determine how people will behave.

Jean Piaget, the Swiss psychologist whose studies on the development of thinking in children have been very influential, held that the development of infants' understanding of their environment was a result of their active manipulations and explorations of objects, and that they constructed reality through converging lines of sensory and motor information. One source of their understanding of causes came from the infants' own actions: the actual experience of producing a movement plays a key role.

Arm movements made by newborn babies are easily dismissed as unintentional, purposeless, or reflexive. Spontaneous arm-waving movements were recorded while newborns lay supine facing to one side. They were allowed to see either only the arm they were facing, or only the opposite arm on a video monitor, or neither arm. Small forces pulled on their wrists in the direction of the toes. The babies opposed the perturbing force so as to keep an arm up and moving normally, but only when they could see the arm, either directly or on the video monitor. The findings indicate that newborns can purposely control their arm move-ments in the face of external forces, and that the development of visual control of arm movement is under way soon after birth. Newborn babies purposely move their hands to the extent that they will counteract external forces applied to their wrists so as to keep the hands in their field of view.

While watching their moving arms, newborn babies acquire important information about themselves and the world they move in – information babies need for later successful reaching and grasping, from around four months. To successfully direct behaviour in the environment, the infant needs to establish a bodily frame of reference for action. The child, for example, needs to learn about its bodily dimensions in relation to its movements. Here vision plays an important role, and many lessons have to be learned in the early weeks before reaching for toys can emerge. Babies have to learn, for example, how long their arms are in order to perceive what is within reach and what is out of reach. It seems likely that a fast-growing child will need constantly to recalibrate the system controlling movement, both visual information and innate knowledge of the position of the limbs and the body.

Babies learn that their own movements can cause motion. They can pull a string attached to a mobile. If a ribbon is tied to the baby's foot and the other end to a mobile, they rapidly learn to kick and so make the mobile turn; a week later they will remember how to do it. But when you disconnect the ribbon they continue to kick. Piaget called this a magical action, as they coo and smile at the mobile at the same time in order, probably, he thought, to get it to move. They are acquiring causal beliefs. Piaget's studies on children led him to the conclusion that at an early stage in their development they had what he called feelings of participation in natural events that were accompanied by magical beliefs. The child's movement makes the sun move, and the wind can obey them. Who, they wonder, is in fact pushing the wind? Do the clouds make the wind? Piaget thought that explanation of movement is the central point to which all the child's ideas about the world converge. Influential as these ideas about children's magical explanations have been, more recent evidence offers rather little support for them.

Experiments have shown that babies just a few months old already perceive the world as being composed of cohesive solid bodies that keep much the same form when stationary as when moving. In addition, they have a special system in their brains – a module perhaps – for mapping the 'energy' of these objects, some measure of their mechanical properties that can be likened to the concept of 'force'. This concept gradually develops and is consistently present at two to three years old. Already at this early age, children know that a moving object – a ball – can make another move on impact. This primitive concept of mechanics may be the key causal belief that originally evolved in early humans, for, as I shall argue, it was essential for making complex tools.

Studying young children's causal beliefs makes use of the standard habituation and dishabituation technique. This technique uses the fact that babies like novelty and will look longer at new things: the same thing over and over again causes them to habituate and lose interest, while a new thing causes them to regain their interest, to dishabituate. This technique is widely used to test babies' beliefs: they look longer at novel or surprising events involving objects that go against normal causes, like a box at the edge of a shelf not falling down, or an animal passing across a window without being seen. Babies already apparently have clear concepts about physical causality. They perceive physical events according to three principles which may be genetically determined: moving objects maintain both connectedness and cohesion, and do not break up or fuse; they move continuously and do not disappear and appear again unless there are other objects in the way; they move together or interact only if they touch. There are many experiments to support this. For example, they clearly expect that for a moving block to make another one move, it must make contact with it. By eighteen months, babies will use a 'tool' like a rake to pull a toy towards them that was out of reach; we will see how hard this is for chimpanzees.

Are children actually born with beliefs about objects and their physical properties, as just described, or do they rather have a mechanism that guides their acquisition of knowledge about objects? The latter seems more likely, and it is genetically determined that the brain has the ability to acquire such beliefs. That genes can specify modules in the brain that provide particular ways of thinking is nicely illustrated in relation to numbers. Babies as young as one month can categorise the world in terms of numerosities, and infants a few months old can already add and subtract, as shown using the habituation technique.

Young children thus perceive that certain objects have causal properties with a renewable source of energy or force, and this is a most sophisticated idea unique to humans. These special objects, or agents, can act in pursuit of goals. They also have a concept of 'Force', which is a primitive mechanical notion, not the same as the scientific concept of force. The basic idea is that when bodies move, they possess a force; and this can, on impact, be transmitted to other objects, which can receive or resist. Infants expect a stationary object to be displaced when hit by a moving object, and by six months can reliably estimate how far it should move. It is likely that these key principles are learned. They know a glass disk would break if dropped, but a metal one would not. They are also already aware that the size of an object affects whether it can pass through a gap before they realise that it also affects the size of the container that holds it, or that the size of a bulge under a cloth signals the size of the object underneath. They will remove a pillow placed in the way of a wanted toy. They also have an appreciation of gravity: they expect a toy car rolling down a slope to increase its speed, and if a falling ball suddenly stops in mid-air they are puzzled.

Babies already know about hidden objects if they see them being covered; but even at fifteen months are mystified if you put a small toy in your hand, place it under a cloth where you leave the toy, and then show the baby your empty hand. Infants can imitate, copy motor acts spontaneously, and they can repeat the movements carried out on novel objects reliably one week later. A nice example of babies' concept of cause and effect is provided by their using a new way to execute a task after having seen an adult do it; they do not simply imitate. If a fourteen-month-old infant sees an adult illuminate a light box by bending over and touching it with the head, they will light the box one week later using either their hand or head. They use their head more often if in the original demonstration the demonstrator's hands were free, but if they were clearly occupied, like holding a blanket, then the infants use their hands more often to turn on the light. Thus, rather than just imitating, they infer that when the hands are free and not used, using the head must provide some advantage.

Babies less than a year old have the ability to retrieve an object by pulling on the cloth on which it is resting. At this early stage they do not use a stick with a hook-like end to get the toy unless it is already placed within the hook. At eighteen months, when they begin to do so, they are effectively using objects as tools, which is a major advance. They have moved from believing that two objects must have a point of contact between them to initiate movement, to using their causal knowledge to put the tool in contact with the object. Two-year-olds understand imagined causal events such as Teddy being wet after a make-believe bath.

Language learning is a key process, and there is evidence that children start very early and learn a great deal about language before they can actually speak. This learning has a major social component. Language is important for acquiring many causal beliefs. There can be no doubt that language learning involves key genetically determined features in the brain, which are unique to humans.

As early as one day old, babies discriminate between happy and sad faces, and by six months, between happy and angry voices. At this age they manipulate objects and interact with people, and then at around one year old they exhibit joint attention, in which they interact with both people and objects at the same time. This involves pointing, imitation, and imperative gestures, and almost all infants show these skills at this age. It is clear that the acquisition of these skills, or the ability to acquire them, is genetically determined, that is the infant's brain is 'designed' for the purpose of social interaction and understanding. Evolution has resulted in the selection of that design, as it is a great advantage to the individual to interact with, and understand others (children suffering from autism have significant problems with these skills related to joint attention). This transition may be fundamental to human social development, as it is an attempt to comprehend others by applying what they have themselves experienced. A new understanding now emerges: an early set of beliefs. A key belief at this age is that other people are 'like me'.

Very young children are aware that people have different desires, and by eighteen months understand that they have different beliefs. A three-year-old child grasps the causal basis of its own actions, and from this then develops an understanding of others. Children initially explain people's behaviour in terms of their own feelings and desires. Two- and three-year-olds can tell lies, and this means that they need to appreciate the difference between what they and someone else believes. They need to understand belief. And they can be terribly incompetent liars, because they are not yet fully aware of what it takes to make someone have a false belief. 'I did not cross the street by myself' the three-year-old shouts from the other side. By the age of four they understand that other people not only have intentions, but beliefs, which may or may not be expressed but which will affect how they behave. A little later they can say, 'Does she think like X?'

Young children already distinguish between people and inanimate objects, and are sensitive to the differences between human action and the motion of objects. They know that two inanimate objects have to contact each other if a causal event is to take place, but that with people, a causal event does not require physical contact. Babies one year old already point at things, which is something no ape, young or adult, ever does. Babies do this to get a toy before they can talk. It means that they know that what they see, some other person can also see. They

41

reason about human action within the first year, and this differs from beliefs about moving objects. People, unlike objects, do not act according to the contact principle; they are self-propelled. Do children actually suspend the contact principle with events involving people? The answer is yes; and they understand there are communicative interactions between people.

The distinction between animate and inanimate is an important feature of children's development. They interpret an object's composition and pattern of movement in ways that are relevant to understanding the distinction between 'animals' and 'non-living' objects. The key is that animate objects can cause themselves to move or change their form, whereas inanimate ones cannot. Understanding that distinction requires a concept of cause.

Learning about animate motions requires an insight into where the energy for the movement comes from. For objects to move, there must be a force, and unless it comes from within, as in animals, this must be an external force. All this fundamentally requires belief and appreciation of cause and effect. However, the motion of an object alone cannot determine whether or not it is alive. A sudden change in direction with no contact with another object might be interpreted as animate, but it may in fact be due to unseen external forces like a strong wind. Three-year-olds can tell which strange objects could go uphill. A group of three-year-old children were shown a set of pictures of a variety of 'animals' and machines that were totally unfamiliar, but the children were very good at inferring whether or not they could go up a hill by themselves: statues with feet could not do so. They were using their causal beliefs.

There is in the standard developmental sequence of children what is called a teleological stance, a tendency to view objects as being designed for a purpose. So children will judge that an otter spends time in the water because of its webbed feet, even though it looks like a land-living weasel. It was Piaget who first suggested that teleological thought was inherent in childhood and that children see nature as goal-directed. This is common in adults, who will ask why something exists, what its future might be, its properties, or more generally, what it is for. It may be something we are compelled to do because our minds are made that way. It is indeed a basic feature of cognition, particularly related to living things, and may be linked to a specialised module for classifying organisms. A child might spontaneously ask about the function of a pointy part of an insect, but not about a pointy part of a rock. Children at six to seven years prefer a functional explanation for why plants are green, and a physical one for emeralds. For plants they understand that it helps them grow.

A need to explain events is at the core of a child's development, and is as important as the drive for sex or food. They, and we adults, want to understand what is happening in the world around us. This drive consumes children in their first three years and their exploratory drive can be quite troublesome for parents. Across all languages, when children begin to talk, the most common topics are the presence of people or objects, the exchange of possessions, the movement of people and objects, and the activity and intentions of people. Almost all of these involve either peoples' intentions or causal events. Later, their stories will be full of causal and intentional links. In all languages, the concept of causality plays an important structuring role – causality is a fundamental aspect of human cognition. The statement 'You broke the glass' is causative. Quite different sorts of explanations relate to function. Children never pick up a rock or some sand and ask 'What is this for?'

When children observe objects such as self-propelled balls interacting, they interpret the interaction in terms of positive and negative values. For example, if one of the balls comes into contact with another and immediately moves away, and the other ball has its shape altered, this is seen as hitting. If the ball comes gently into contact and moves back and forward on to the other, it is perceived as a caress, and they expect the target ball to respond in an appropriate manner. Children thus develop a set of ideas relating these concepts to social events. They understand helping and hurting, and can interpret the behaviour of balls in such terms, one ball crossing another, or helping it out of a confined space. Infants expect positive acts to be reciprocated. All these ideas are part of a theory of mind, where explanations are put forward in terms of how other people are thinking.

Questions asked by pre-school children provide insights into how they think about cause and effect. Even before three years old, toddlers talk about causes with surprising sophistication. The earliest questions relate to the social rather than the physical world. Quite often questions arise in situations where their expectations were violated or unexpected. Typical among questions are those that ask how – how do they make statues; why – why does it rain sometimes; what if – what if someone's head were cut off; what for – what is this stick (the gearstick) for? 'How' questions increase in the fourth year, as does interest in biological phenomena, whereas interest in physical phenomena decreases. Typical questions are then: why can we see the stars? how do people die?

Thus, by the age of four, children have a well-developed theory of mind, and recognise that others have an image of the world, and beliefs that are different from real objects, and that these beliefs determine to a large extent how people behave. At a somewhat older age they recognise traits: that individuals behave in a particular way by virtue of the kind of people they are. Thus they may think that a person may be lazy, or cruel, or unreliable, or that it is in the nature of dogs to bark. These ideas can also have a strong racist content that they have learned: that

Jews are mean, blacks are lazy, and so on, are not uncommon stereotypes.

What is seen can influence what is believed. In one study, children were shown two containers; a whole small cake was placed in one, while only a small piece of cake was placed in the other. When asked which container they preferred, they all chose the container with the whole cake. They were then told that another child would have first choice, but they could not observe what that child chose. So which container would they then choose? Almost half of four-year-olds chose the empty container. They could infer that the other child would take the whole cake, but this belief could not overcome their 'seeing' the whole cake in the container they chose.

A test for whether children have a theory of mind, that is, whether they can understand what others are thinking, involves showing them a small tube with a very characteristic pattern that normally contains Smarties: the well-known sweets. A child, when asked what such a tube contains, will say that it contains Smarties. The child is then shown that it does in fact contain pencils and is then asked what their best friend would think was in the tube. Children with a theory of mind will reply that the friend would think it contained Smarties. Autistic children cannot give the correct reply, and will suggest pencils.

Most species react to similar stimuli in a similar way, like the frog shooting out its tongue at any appropriately moving object. By contrast, humans try to distinguish between such similarities, and try to understand what the cause is. Children are aroused by objects that move on their own, and they reason about their goals or intentions. This can be illustrated by an experiment in which they had to judge what caused a Jack-in-the-box to pop up. Two events arrived at the base of the box at the same time – a rolling ball and a series of lights – which seemed to influence the pop-up. Children judged the ball to be the cause. But when these two possible causes were separated from the box by a short distance, the light was judged to be the more likely cause. The children clearly had some knowledge of causal mechanisms at this simple physical level.

Children can distinguish between physical objects and an imagined one. They know you cannot touch an imagined piece of cake. While children can distinguish between what they imagine and what is real, as many as a quarter of four-year-old children have invisible friends or invisible animals with whom they talk quite happily, and who have a certain reality. There's a nice experiment in which someone tells a story and says, 'Now look, here's this little house and there are two little holes in it and in this one there's a puppy dog and in this one there's a monster, which one do you want to feed?' On the whole, they don't mind putting their hand into the hole that has the puppy dog. I think I'd do exactly the same. It's a non-trivial matter to link one's beliefs reliably to the real world.

To what extent is children's understanding of the mechanics of objects reliable? They have difficulty thinking about chance outcomes; they show surprise when witnessing events with no obvious cause. Magic shows demonstrate the complicated physics of everyday life that the tricks contradict: things change into other things, appear, disappear. In a quite complicated experiment, a postage stamp was put in a box and then after external manipulations, like cutting another stamp in half, it was itself found to have been cut in half. Children attributed this to the experimenter's action. Six- to nine-year-old children accept a mechanistic explanation, but some invoke a magical explanation. They do acknowledge magical outcomes as a special class of phenomena, and this usually occurs when they are faced with puzzling processes. Children give magical interpretations for extraordinary events, like a toy car changing colour when dipped in cold water. They may also prefer magical interpretations if using them avoids a loss of some sort, or gets them out of a dangerous situation. They also accept magical transformations in fairy tales.

It is only from around four years that children use the concept of quantity, including numbers, properly. There are lots of studies going back to Piaget about the difficulty children have at a certain age with conservation of volume. For example, I have a certain volume of liquid, which I put into a long thin glass, and then put exactly the same amount of liquid into a short fat glass. Which glass, I then ask the children, has more liquid? They initially always say the tall glass, a common-sense belief in a way, since the height of the glass makes them think so. It takes them quite a long time to realise that it doesn't matter into what glass the liquid is poured. They only understand conservation of volume at a later age.

When we come to consider how children look at the real world, there is a temptation for some developmental psychologists to see them as little scientists. But even secondary school children have quite a lot of difficulty with quite simple scientific concepts. For example, they think that when an object is moved higher it becomes heavier. But, as we shall see, causes in science go against common sense. And children will provide nice examples of how they distort their own observations. They will cook their observations in order to maintain consistency with their beliefs. Studies show that if you take a heavy lead weight and put it on a sponge and then on the table, children say they saw the table go down then come back up again. The relationship between perception and cause is not simple. Five-year-olds understand that fan A cannot blow out a candle because a shield is in the way, but that when, after five seconds, fan B is turned on and the shield moved in front of it, it is fan A that blows out the candle. However, only by nine or ten do children understand the causal mechanical principles that make simple machines work.

There is quite widespread belief among children that illness is a punishment for wrongdoing, and that they are to blame for their illness, but there are also beliefs about germs and food being causes. These beliefs fit with Piagetian stages: pre-logical, concrete logical, and formal logical. The first is associated with illness having an external cause – the sun gives one a cold, God gives one measles. At later stages, contamination is a characteristic belief. Young children who explain that one gets AIDS from sex have no idea why. And it is very interesting to look at children's beliefs in respect to their own illness. At an early stage, they assign it to mystical forces. My grandchild, at the age of eight, thought illness was a punishment for being naughty. As regards death, there is some evidence that some children view death as reversible, and even personify death as a kind of bogeyman.

Pre-school children do not regard reproduction as a key feature characterising animals. Do pre-school children have biological knowledge relating to growth, inheritance and infectious diseases? When asked where babies come from, some three- to four-year-olds see it as a geographical question – you get them from hospitals, or buy one. Others think of it in terms of making the baby – the mother swallowing something and making it in her tummy. A baby pig grows into a big pig, not a cow. But reproduction is not a core concept, and nor is contagion. Culture also affects children's beliefs. Western children say they will die if they do not breathe. Japanese children construct a vitalistic biology. They eat to obtain vital power from the food. In their beliefs a vital force, a life force Ki, plays a central role; and when a person or animal dies, this vital force leaves the body. Air is one source of this energy.

Counterfactual thinking, thinking about the result if some

event had not occurred, contributes to causal thinking. Young children compare an observed outcome with what has occurred in different circumstances; if no one had been blowing the candle it would not have gone out. Words like 'nearly' and 'almost' characterise such thinking, and these words are used by children before their third birthday. A little later they can imagine a different set of causal events. Sally would not have got dirty fingers if she had chosen a pencil and not a pen to draw with. In general, both children and adults use counterfactual thinking in connection with an unpleasant outcome; they think back to see how it could have been avoided.

Many causal beliefs come from parents, teachers and friends. Key religious beliefs come in this way, and build on – but also violate – ordinary causal ideas. In many religions there are special beings who hear and receive messages and are also, for example, able to read our minds and pass through solid barriers, and are immortal. Children have a human-like conception of God. They also accept that God, unlike ordinary humans, would know what was inside a closed box without having to open it. American six-year-olds strongly endorse the idea that animals, as well as objects, are made by God.

The evidence that having causal beliefs is innate should now be very clear. These beliefs are about both the physical and the social world, and are related to the child's experiences. Children begin to realise that correlation and coincidence do not necessarily underlie causality without an idea of mechanism. Thus, night does not cause day. If children, and even adults, cannot figure out how an event comes about, neither doubts that there is a cause, even if we do not know the mechanism. Children assume there is a mechanism, and will search for it.

All these causal beliefs in children are determined by our brains giving us the ability, from a very early age, to have the concept of cause. This ability may require special modules that have evolved by changes in the genes that control the development of the human brain. Over millions of years these changes, together with those in our bodies, have made us fundamentally different from all other animals. Why did this occur, and why was the acquisition of causal beliefs so important? It is all related to tools, which may sound improbable, but I hope to persuade you. First, however, we should look at animals, and examine if they have beliefs similar to children.

CHAPTER FOUR

Animals

> ... humans, but no other primates, understand the causal and
> intentional relations that hold among external entities.
>
> Michael Tomasello

To what extent do animals have causal beliefs – that is, beliefs about what causes movement and events in the world around them? The answer is rather little. It is this causal understanding that makes us human. For humans, the weight of a falling rock 'forces' a log to splinter, and wind makes the tree's branches shake. No animal has a similar understanding of force, or cause and effect. One may illustrate the differences in chimpanzee and human thinking with the claim that the apes, seeing the wind blowing and shaking a branch till the fruit falls, would never believe, from this, that they could shake the branch to get the fruit. Nevertheless, they are among the animals that have simple causal beliefs, but unlike humans, they never make complex tools. The field is, however, a controversial one. In spite of the evidence that primates have very limited causal beliefs, many workers in the field still believe that primates do have causal beliefs that are not all that different from human children.

Darwin was very insistent on the continuity of mental skills from primate to humans, and for example observed that the origin of tool use could be seen in chimpanzees cracking nuts. But he was forced to concede to the Duke of Argyll, who claimed that 'the fashioning of an implement for a special purpose is absolutely peculiar to humans'. Both Povinelli and Tomasello have put forward the key idea I am presenting here, namely that animals other than humans do not have causal beliefs – they do not have explanations as to why objects interact with each other in the ways that they do. They have both shown that while some primates, like chimpanzees, can perceive and interact with their environment in a similar way to humans, they do not have concepts of different causes to explain interaction between objects. Tomasello is quite specific when he states that non-human primates do not understand the world in causal terms. They may learn that one event leads regularly to another, but do not understand the causal forces mediating these relations. Povinelli points out that when primates like chimpanzees use simple tools, it mostly involves learned procedures, and only a very limited concept of causation. He also explains that understanding why events occur opens the way to technology. But some animals are at the beginning of causal thinking, as we shall see.

Mammals, particularly primates, are very good at learning complex tasks, and often seem to show causal understanding. I am sure that many pet owners believe that their cat or dog has some causal understanding. And there are thousands of experiments with rats being trained to press levers and carry out quite special movements to obtain food. These learned tasks do not reflect any understanding by the animal of what it is actually doing. As described below, there are studies showing the absence of true causal reasoning in animals, though there are cases where it seems that basic elements of the process might be present. It may be helpful to distinguish, again, between weak and strong causal knowledge. Weak causal knowledge is the result of associative learning – one event is frequently followed by another one – and usually many repetitions are necessary, as is the case when a rat learns to press a lever for the reward of food. By contrast, strong causal knowledge is based on interpretation, and may relate to events widely separated in time or space: we believe, for example, that damage to the car brakes could later lead to an accident. Animals do not have causal beliefs of this type, though there are challenges to this view, namely that there is a concordance between human causal judgements and, for instance, a rat's goal-directed actions. For example, the actions of rats in pressing a lever to obtain a different food from that which had made them ill on a previous occasion cannot, it is claimed, be explained by stimulus-reinforcement learning.

Humans, even children, as we have seen, have an immediate sense of cause when they see a moving object like a ball hitting another ball. They have an intuitive concept of force. This concept of cause is different from that learned by the association of events. Repetition of a pair of events, such as a whistle followed by a bell, can lead one to believe that the former caused the latter. This type of learned association is not a 'belief' in animals, as has been claimed. In one experiment, chicks were fed in a 'looking-glass' world: when they moved towards the bowl of food, it retreated, but it came towards them if they moved away from it. They never adapted to this situation, for they failed to appreciate the causal connection.

There are important cognitive similarities between humans and mammals, especially primates, who remember their local environment, take novel detours, follow object movement, recognise similarities, and have some insight into problem solving. Primates recognise other individuals, predict their behaviour, and form alliances. They can also differentiate between animate and inanimate objects. However, they have a limited understanding of the intentionality of other animals, and only a glimmer of understanding of the causal relationships between inanimate objects. They do not view the world in terms of the underlying 'forces' that are fundamental to human thinking. It is not that chimpanzees lack visual imagination or are unable to learn quite complex tasks by trial and error, but they do not reason about things. They have, for example, no concept of force, and even more important, no concept of causality.

Some experiments on chimpanzees show that they do appreciate that contact is necessary in using a tool to get food, but will focus only on the contact and not the force it generates on the target object. A hook at the end of the stick is not perceived as a means to get the reward compared to one that is straight; only contact is recognised as relevant. They do not understand the world in intentional or causal terms, nor understand the causal relation between their acts and the outcomes they experience. Apes, for example, cannot select an appropriate tool for a simple physical manipulation without training. Nevertheless, tamarin monkeys are able to correctly choose the right simple tool to get food.

Evidence for these claims comes largely from studies on the folk physics of chimpanzees by Povinelli. Folk physics is used to describe the way people use common sense to think about how the physical world works. It is almost always different from scientific physics, for the world is not built on common-sense notions; scientific beliefs are in this sense unnatural, as will be discussed later. 'What sort of folk physics do apes have?' is the question posed by Povinelli. He has carried out experiments that show that while many of the abilities of humans and other primates to perceive and move are similar, primates like chimpanzees do not have concepts of different causes to explain interaction between objects.

One might have thought that Wolfgang Kohler's classic experiments with chimpanzees showed just the opposite. His studies on chimpanzees, some eighty years ago, showed that they could sometimes, perhaps with some training, stack boxes on top of each other to get a banana nailed to the ceiling. But Kohler himself acknowledged that the chimpanzees had no knowledge of the forces involved. For example, they would try to place one box on another along its diagonal edge; and more strikingly, if stones were placed on the ground so that the box toppled over, they never removed the stones, and so never reached the bananas.

In an experiment by Povinelli's group, apes could choose one of two rake tools to obtain a food reward. The choice was between dragging food along a solid surface, or dragging it over a large hole into which the food would fall and be lost. Only one of six apes was successful in this test at the first trial, and this solitary success may have been due to chance, although the apes did eventually learn by trial and error. They also did badly with an inverted two-prong rake that could not move the food, and on tests with flimsy tools. Again, when required to get a banana by pulling on a rope, they could not distinguish between a rope just lying on, or merely being very close to the banana, and a rope that was actually tied to the banana. They showed no appreciation of physical connection as distinct from mere contact. It should be recognised that there has been criticism of Povinelli's work. Chimpanzees' skills, it is said, only reveal themselves in adulthood – when they are eight to nine years old – but Povinelli often used five- and six-year-olds. Moreover, some of his results have not been peer-reviewed and so not published in standard journals. There are also criticisms about how his apes were reared.

In another series of key experiments by Povinelli, chimpanzees were set the task of using a stick to push food out of a clear tube. The tools were of various sizes, some being too short, too thick, or too flexible. An understanding of basic forces should enable an individual to choose the right tool. They could do the task, but only after much trial and error. In another test there was a small trap under part of the tube, and to get the food the subject needed to push the food from the end of the tube that avoided the trap. In over seventy trials, the chimpanzees' results were no better than what they would have achieved by chance. Then, eventually, when the animals had learned to do it, the tube was rotated through 180 degrees, so the trap was now on top and had no effect on getting the food. But they continued to push the food away from the trap. By contrast, two- to three-year-old children understood what to do from the earliest trials.

In experiments to distinguish between adequate and inadequate support of an object using looking-time measures, the chimpanzees' sensitivity to different support relations between two objects was assessed. In each experiment, the chimpanzees saw a possible and an impossible test event, presented as digital video clips. Looking times suggest that chimpanzees use the amount of contact between two objects, but not the type of contact, to distinguish between adequate and inadequate support relations. These results indicate that chimpanzees have some intuition about support phenomena, but their sensitivity to relational object properties is not as good as that of human infants.

Similar lack of causal beliefs is shown by reports on the behaviour of two male macaques who began using sticks to rake apples fallen from a tree outside the fence enclosing them. They tried to reach them by hand, and would only grasp stones and shrubs, which they released. Occasionally, they grasped two sticks, which they rubbed together, or threw sticks and stones at the apples. Eventually they used the stick on the apples, but pushed them away as often as pulling them closer. Young children understand the correct principle after three trials, while the macaques took fifty sessions of thirty minutes each.

Do any animals have any causal understanding when they

use simple tools? In its simplest form, a tool is used for some very basic essential purpose, such as acquiring food. A wide variety of animals use simple tools, and even modify them. Finches in the Galapagos use a cactus spine to probe for termites; the Egyptian vulture drops rocks on to ostrich eggs to break them; other birds use stones to crack open clams; the mud wasp holds a tiny pebble in its jaws to tamp down mud in nest building. One captive black-breasted buzzard that dropped stones on domestic hens' eggs preferred a 40 g stone from a range of stones weighing 15–65 g. But there is no evidence that in these cases there is any causal understanding.

Tool manufacture, implying substantial modification of some material to produce a tool, is rare, but can be routinely found in two genera of primates, and several birds. Natural tools used by apes are sticks or stones, but a tool modified intentionally is rather rare, though chimpanzees do show some evidence for this ability by trimming twigs to dig out insects. Modification of tools thus does occur, but there are no examples among animals of combining two elements to make a tool.

A remarkable case of tool use is found amongst New Caledonian crows. Only twenty-six of an estimated 8,600 known species of birds have ever been shown to use any kind of tool, and in many of those cases, only a small fraction of individual birds do so. New Caledonian crows have only recently been the object of scientific study, but their use of tools in the wild appears to occur with sufficient frequency that local people often comment upon the abilities of the crows. Their behaviour has recently been used as the motif of a postage stamp from New Caledonia.

New Caledonian crows manufacture and use several types of tools for extractive foraging of invertebrates, including straight and hooked sticks, and complex stepped-cut flat tools made from leaves. These tools have some of the hallmarks of complex tool manufacture; form is imposed on the raw material with control of various shapes. A skilled tool-making technique is involved, and there is even standardisation of the shape of the finished tools.

Further impressive evidence of their skills comes from the crows being given a specific task. The task given the crows consisted of extracting food from a transparent section of horizontal pipe. Ten sticks of increasing length were cut from bamboo and displayed in length order. The crows were capable of selecting the stick that matched the distance to the food significantly more often than would be expected at random. Choosing the longest tool also appeared to be an important strategy. This ability of New Caledonian crows to select an appropriate tool on their first exposure to a task that is novel to them is impressive. Moreover, this ability seems to be inherited, as naive juvenile crows can make these tools without contact with adults. Another bird with apparently causal skills is the raven. Ravens confronted with the problem of food dangled on a string were successful in pulling up the meat by pulling up the string in a few minutes, and this suggests some possible causal insight, as it was not based on trial and error learning.

There is also evidence for some causal understanding in primates. Monkeys and chimpanzees place thick-skinned or armoured fruits on an anvil of stone and smash them open with another stone or a heavy branch. It can take them up to ten years to acquire the necessary skill, and there is no evidence of them modifying the stones. They do nevertheless leave their 'hammers' behind near the fruit trees and return to use them the following day. Chimpanzees can climb fruit trees with long sharp thorns by ripping off the bark from a tree and using pieces as sandals to protect their feet. At a height where there is much fruit, a female may take some bark to use as a comfortable seat.

In the tropical forests of West Africa, chimpanzees have been observed spending hours using stone or wooden hammers to break open the shells of nuts by first placing them on a stone, the anvil. Moreover, this behaviour is ancient. A chimp nut-cracking site has been discovered on the Ivory Coast, where numerous stone pieces have been dug up. They were, unlike hominid tools, created by accident. They do have some resemblance to early hominid tools going back more than 2 million years. But it is relevant to my argument that these sites show no evidence of real tool making, or the selection of stones with respect to the material of which they are composed; only the weight of the stone appears to have been selected. However, it has been observed that the stone on which the nuts were being cracked, the anvil, was made level by inserting a stone under it at the low end. This is an example of a physical causal understanding.

Chimpanzees and apes are thus at the edge of causal understanding as shown by their use of simple tools, such as trimming a grass reed to dig out ants. But in no case of stone tool use is there evidence of modifying the structure of the stone to improve its function. They lack the causal beliefs to transform physical objects into useful tools. This makes sense, as the evolution of causal belief in humans would have been a gradual process, building on skills of the type possessed by those other ancient primates. While there is no persuasive evidence for causal beliefs in primates, there are significant examples that hint clearly that they really are at the edge, even just over the edge. There is an account, for example, of a mother chimpanzee forming a bridge across a chasm by holding on to branches so that her screaming daughter could cross over to safety.

There are other examples that suggest that apes can come close to having some causal understanding, particularly in relation to tool use, when trained by humans. One must always keep in mind that evolution is gradual and builds on, or tinkers with whatever is already there, so it should be no surprise that, with special training, a chimpanzee can carry out human-like activities. Motor imitation of a novel movement is clear in humans, a little evident in apes, but absent in other animals. But the chimpanzee Kanzi, a bonobo ape, showed remarkable skills. He learned from his trainers to create and use stone tools to gain access to food by cutting a rope. He could make stone flakes and evaluate them after observing a human striking two rocks together. On his own, he created flakes by throwing one rock onto another on the ground, suggesting that he may indeed have had a concept of force.

Throwing is certainly an important human activity, whether in recreation or in anger. It is not, however, an exclusively human activity, and other primates are reasonably good at it. For example, capuchins, a species of New World monkey found in South and Central America, can throw stones at both moving and stationary targets with reasonable accuracy, and can use both power and precision grips for throwing. They also use throwing as a way of transferring food between social groups, and capuchins proved quite good at throwing stones into a bucket partially filled with either peanut butter or a sweet syrup. The reward for accuracy was being allowed to lick the stone after it landed in the goo. About half the time they threw from an upright, bipedal position, and most of the time they threw overarm.

While human infants only six weeks old will imitate putting out their tongue to one side, imitation by other animals, including primates, is much less convincing, and teaching off-spring by getting them to imitate a particular behaviour is rare. There is little parental training by chimpanzees of their young, in comparison to humans. However, at about ten months, infant chimpanzees first make sure that their mother is looking at them before requesting grooming. Human-reared adult chimpanzees can copy simple motor actions like placing a ball in a bowl, or wiping a lid with a towel. But they are unable to copy movements of the limbs when no contact with an object is made – they could never be taught to dance unless, perhaps, trained from birth. Again, Michael Tomasello argues that chimpanzee gestures are not acquired through imitation. Young chimpanzees thus learn very little from their parents. When the human experimenters taught selected chimpanzees new gestures outside their groups, and then reintroduced them to the groups, the others showed no tendency to copy the new gestures. However, there are descriptions that claim that dolphins and some primates can imitate.

Chimpanzees can recognise themselves when they look in a mirror: they pull faces, pick at their teeth and ears, and explore themselves. This may help them with tool use, as it distinguishes their action from that of the tool. This could have been an early step on the pathway to causal beliefs in humans. Animals are not good at imitating another's action, but they can copy another's choice of object. A hungry pigeon will choose to peck at a red key rather than a blue one to obtain food if it has seen another, trained, pigeon do so. Similarly, rats will copy pressing a rewarding lever. But they do not imitate the actual movement. If the rat has used its nose, or the bird its wing, they will not copy this movement. Chimpanzees, too, are very poor at such tasks, but human children, by contrast, will copy actions closely, as we have seen, and this undoubtedly is an important human ability of relevance to tool making.

There is, however, good evidence for culture, in the sense of socially transmitted behaviour, in chimpanzees and orang-utans. Different groups of orang-utans use sticks to scratch body parts or forage for insects, or use branches as swatters to ward off insects. These are examples of behaviour that are found in only some groups, and imply social learning. Chimpanzees can be taught gestures, and even use them in the wild; examples include throwing some loose material at another chimpanzee, and slapping the ground while looking at another. The cultural diversity among chimpanzees, however, is greater than that so far recorded in any other non-human species – further evidence, perhaps, that the chimp may not be quite the chump that some, including myself, have supposed. Andrew Whiten and his colleagues have examined the evidence from six different chimpanzee communities in Africa, and identified thirty-nine different behavioural patterns, the majority involving the use of objects. The differences between communities could not be attributed to differences in physical or geographic conditions. For example, nut-cracking is found only in the two western communities, and not in the four eastern ones. Chimpanzees at several sites use sticks to dip for ants, but they use different methods at different sites. Different communities of chimpanzees have independently developed such varied tool cultures as probing for termites, ant-dipping, nut-cracking and leaf sponging. But, and this is crucial, the tool use is stereotyped, and in any one community only a few tools are used.

All the above evidence makes clear that while primates and some birds use simple tools, there is an almost total absence of causal beliefs in animals other than humans. In no case of stone tool use is there evidence that individual animals modified the structure of the stone in order to improve the tool's function, though a few cases of anvil stabilisation have in fact been observed. Crows have created tools with a specific form, but we don't know how they think about such processes and to what extent they have a concept of cause and effect. So while animals like crows and monkeys have some understanding of tool use, they have a very limited capacity for refining and combining objects to make better tools. The tools chimpanzees use have a narrow range of functions and there is little evidence that they can think up new functions for the same tool. Compare this with the way humans use a knife for a whole variety of purposes. Another important difference is that chimpanzees are slow to pick up skills from other animals. In essence, chimpanzees lack the technical intelligence needed for manipulating and transforming physical objects. The general consensus seems to be that primates lack causal beliefs.

Do chimpanzees have a theory of mind? Can they attribute causal beliefs to other animals? Primates can have quite complex social behaviour, and possess a number of sophisticated social-cognitive skills. However, they never point or show objects to others, bring others to a new location, or intentionally teach others; but when raised by humans, they may acquire some of these behaviours. Chimpanzees have a fairly sophisticated understanding of what others can see, and can even follow the gaze direction of humans past distracting stimuli and behind barriers to a specific target. They also understand that another individual cannot see something if its view is blocked by a barrier.

However, there is one task involving gaze-following at which chimpanzees and other primates perform poorly. In the so-called object choice task, an experimenter hid a piece of food in one of two opaque containers, and the subject, who did not see where the food was hidden, was allowed to choose only one. Before presenting the subject with the choice, the experimenter gave it a cue indicating the food's location, for example, by looking at, pointing to, tapping on, or placing a marker on the correct container. The majority of primates, as individuals, do not spontaneously perform above chance levels on this task, no matter what the cue. By striking contrast, dogs, but not wolves, are more skilful than great apes at such tasks in which they must read human communicative signals indicating the location of hidden food. Domestic dog puppies only a few weeks old, even those that have had little human contact, show these skills. These findings suggest that during the process of domestication, dogs have been selected for a set of social-cognitive abilities that enable them to recognise human signs in unique ways. Another example comes from horses. The famous case is that of clever Hans, the horse who could apparently do sums, but in fact merely responded to very subtle signs from his trainer.

The Premacks, quoting Sherlock Holmes, say that causal reasoning includes two components: a keen sense of normality, so that it is possible to recognise whether things are or are not normal, like the dog that did not bark; and sufficient knowledge to make some inference as to why things are not normal. They tested for this ability in young chimpanzees by placing them in a quandary. The animals were introduced to a runway that consistently led to food, but the food was occasionally replaced with a rubber snake, and this resulted in a rapid retreat by the chimpanzee. On the next occasion, the animals who had encountered the snake approached with extreme caution. The next step was to present the test animal with another animal who had just repeated the run and thus could be delighted with having got food, or be very agitated at having found not food, but a snake. Could the test animal benefit from the emotional state and behaviour of the animal that had just been down the runway, and so predict what it might find? The state of the animal had no effect whatsoever on the behaviour of the test animal.

There are studies showing that animals, if painstakingly taught, can have something like appreciation and understanding of simple language. There is at least one parrot, named Alex, who can go beyond imitation. Alex has been taught by Irene Pepperberg to use more than 100 words to refer to objects and actions, and can use simple commands and answer simple questions about the locations, shapes, and even number of objects that are shown to him. Back in 1884

Sir Jon Lubbock taught his dog Van the names of objects like 'slipper' and 'bowl', and actions like 'sniff' and 'fetch'. When words were written on pieces of paper, Van was able to make requests by touching the appropriate word. This approach has been extended to dolphins: after being trained to spit at the ball and touch the hoop, they respond correctly to 'hoop touch'. Much more impressive were the results obtained from teaching the bonobo ape, Kanzi. For the first nine years of his life, Kanzi was given spoken commands and corrected when he got it wrong. This ape's comprehension of language was equal to that of an eighteen-month-old child – and for simple commands like 'go to the refrigerator and get some bananas', the ape was rather more advanced. For certain commands, for both the ape and the child, words like 'some', 'to' and 'on' had no real meaning: the previous sentence would be interpreted as 'go refrigerator banana'.

One way to describe Kanzi's ability to communicate is that it constitutes what has been called a protolanguage rather than true language. Protolanguage has at best a primitive syntax, allowing different combinations of words representing objects and actions. It may be reasonable to infer that at least the potential for protolanguage was present in the common ancestor of the great apes (which of course include ourselves), who were lumbering around Africa some 16 million years ago. A naive chimpanzee may need several hundred trials before it learns its first word. Spurred by these observations, the two Americans hit upon the idea of trying to communicate by using manual gestures, loosely based on American Sign Language (ASL). They were able to teach well over a hundred gestures to another young chimpanzee named Washoe.

Apes could be the only primates other than humans that can pass what is known as a false belief test. This test requires that the animal understands that another individual holds a belief that the subject knows, or believes, to be false. Apes are some-what puzzled when a film (in which chimpanzees chase, capture and dismember another monkey) is shown backwards: they look at it longer. But the evidence for apes having a theory of mind is far from conclusive.

The chimpanzee Sarah was shown films of an actor strug-gling to solve various kinds of problems, like trying to reach a bunch of bananas out of reach over his head, or getting access to bananas when blocked by a large box full of bricks. On other videos she saw him dealing with a disconnected hose, or a gas heater that was not lit. She was then shown one of these films, and given an envelope containing a number of large photographs that could provide the solution to one of the problems, for exam-ple the man getting onto a chair to reach the banana overhead. Her choices were remarkably good. She did make a mistake with the box preventing access to the fruit: she chose to push the box, instead of first removing the bricks inside it.

While having little understanding of the motion of objects, many animals have amazing abilities to control their own movements. Monkeys can do extraordinary jumps from branch to branch, and birds navigate with great skill. Just recall the precision with which the tiny fly lands on the edge of your glass. But these movements do not involve an understanding of the movement of physical objects, or physical cause and effect. Comparative studies of chimpanzee tool use indicate that criti-cal differences are likely to be found in mechanisms involved in causal reasoning rather than those involved in simpler move-ments. Available evidence implies that higher-level perceptual areas of the brain are required for causal thinking.

There can be no doubt that human cognition, emotion, and beliefs have evolved from those of some ancestral primate who probably had similar brain functions to modern primates. Evolu-tion never creates anything totally new, but tinkers and modifies what is already there. For example, our limbs evolved from the fins of ancient fish and our jaws from the modification of gill slits. A particularly striking example is one of the little bones in our middle ear, the stapes, which was originally a bone in the lower jaw.

It has been proposed that just as understanding of relationships in primates evolved first in the social domain to comprehend third-party relationships, so human causal understanding evolved first in the social domain so that the intentions of other humans could be understood. This allowed them to predict and explain the behaviour of others. The advantage of this understanding, it is claimed, is to enable humans to solve problems in a creative way and lead to improved tool making and use. But this, for me, is quite the wrong way round, since I argue that it was causal thinking that was a fundamental adaptation required for making complex tools, and that it was technology that drove human evolution, as will be discussed next. It is not social relations or understanding that led to tool making, hunting, or agriculture. It is hard to see what evolutionary advantage further social interaction could give, or how it could lead to tool use. And just consider the difference in motor skills between humans and primates. Primates can climb, jump and run with great skill, but they have nothing like the skill of a concert pianist who remembers a Beethoven concerto, and whose fingers perform most astonishing and precise movements. It is these motor skills that make us human, together with causal understanding and cultural transmission. There is very little cumulative cultural evolution in any animals other than humans, with the rare exceptions found in some primates.

Tools and causal beliefs may be the basis of the human fascination with ball games that include billiards, soccer, tennis, bowls, cricket and baseball. All involve focusing on how a ball will behave when struck or thrown, and thus involve a basic understanding of the physical forces involved. For these games we use just the same sets of beliefs that were fundamental to the causal beliefs that evolved for tool making. Perhaps all these sports reflect an innate interest in such processes, especially for those no longer making or using tools. No animal other than a human could play golf or billiards.

Thus we can conclude that while non-human primates have an understanding of all kinds of quite complex physical and social concepts, and can distinguish the animate from the inanimate, they do not view the world in terms of the kinds of intermediate and often hidden 'forces' – underlying causes, reasons, intentions, and explanations – that are so important for human thinking. But the beginning of causal beliefs was there in ancestral apes, and it was from this that our causal and technological skills evolved. It is the technological path that we humans took that has separated us most profoundly from our primate ancestry, and from our extant primate relatives. Our technological adaptation has been shaping our evolutionary trajectory in crucial ways for the past several million years. Note how different this view, to which I am totally committed, is from that who put the emphasis on social relations. If you were to go into the jungle, which would you prefer to have with you, a friend or an axe?

CHAPTER FIVE

Tools

... it is evident that man may be distinguished
as the tool-making primate ...
Kenneth Oakley

At the start of *2001: A Space Odyssey* a chimpanzee suddenly realises the awesome power of a basic tool, picks up a stick, uses it as a weapon, and so dominates all his companions. This is a key to understanding the origins of causal belief and tool use to change the environment, defend, and attack. No animal other than a human being has ever properly realised this, though there is some evidence that wild chimpanzees use sticks and stones as weapons against other males or other apes like baboons or humans. And while some primates, as we have seen, do use tools, they have no clear understanding of cause and effect, or the differences between tools and how they may be used in a variety of ways. No animal uses a container to carry, for example, food or water; pots and bags are totally human, and were invaluable for our ancestors when travelling.

But which served as the prime mover in the evolution of the human brain: technology or social behaviour? And what were the adaptive advantages that led to the evolution of the brain so that it acquired causal beliefs? What is the relationship between language, tool use and causal beliefs? Could there have been a mutual positive feedback between all three?

Many common activities have been crucial to human survival

– getting food, staying safe, and reproducing – almost all of which require a concept of cause and effect. There are 'simple' tools, like digging sticks, that humans use in a complex way. Humans will also spend hours tracking game. Monkeys and apes have considerable cognitive skills, which they use when competing for resources; they have a dominant hierarchy, long friendships and social grooming; they indulge in deception and are aware of the personal characteristics of others. But only humans effectively cause one object to interact with another, or with the environment, in a multitude of different ways, and invented technology, which drove human evolution.

All this, I suggest is due to the evolution of the ability of humans to have causal beliefs, which, as we have seen, is a developmental primitive in children. Tool use requires a deep understanding of the physics of the human body, as well as that of the surrounding objects that interact, but it has received relatively little attention from mainstream neuroscience. Children at eighteen months can formulate motor plans related to simple tool use. Some of our understanding of the processes involved come from patients with brain injury suffering from apraxia. For example, a patient might attempt to brush his teeth with a comb, or cannot act out how a familiar tool could be used. Specific regions of the brain have been identified with tool use, and this should help to identify the evolutionary changes in the human brain that can account for it.

It is not clear what motor control systems and other brain systems involving causal belief are required in order to make a tool. It presumably requires imagining the result, and of course it can be learned from others. It has been suggested that a key component of this ability is a causal operator in the brain, which may involve connections between the left frontal lobe and left orientation area. There is evidence from brain imaging studies that distinct brain regions are related to knowledge about different classes of objects, such as people, plants and tools. That tools should relate to a specific brain area again emphasises their importance in our minds. Patients with strokes who have damage to these areas have great difficulty with causal thinking, and often do not know why something happened. It may be a fundamental aspect of our brain to see events in causal terms. The evolutionary origin of such beliefs is, I believe, linked to the advantages to humans of making tools and using them to modify their environment to promote their survival.

As long ago as 1927, de Laguna doubted if complex tool making, which requires planning, could have occurred without language. Others, too, argued that the first language consisted of manual signs imitating the operations of tool use – vocal expression may have come later. But as almost always, there is a striking lack of thinking about causal beliefs as a key mechanism. It is the Premacks who are among the very few who ask the question about how the concept of cause might have evolved. One possibility, they suggest, is that causal belief evolved in the context of personal action. They seem to prefer its origin, like many others, to be in the social domain, but are quite unclear as to how this occurred; perhaps the advantages to the individuals in terms of causal understanding in the social world may have been based on cooperation in using tools. But this implies that tool use arose independently. The relationship of tool use to language will be considered below.

It was Kenneth Oakley in 1949 who wrote 'Modern civilisation owes its form to machine-tools, driven by mechanical energy; yet these perform in complicated ways and use only the same basic operator as the simple equipment in the tool-bag of Stone Age man: percussion, cutting, scraping, piercing, shearing, and moulding.' He also made clear that the men who made tools such as Acheulian hand axes must have been capable of forming in their mind images of what they were trying to achieve. 'Human culture in all its diversity is the outcome of this capacity for conceptual thinking . . .' This original idea of Oakley is at the core of this book.

Even though human ancestors were on two feet and had free hands some 4 million years ago, tool use is usually assumed to only go back 2 to 3 million years. However, it has been pointed out that this fails to recognise that the evidence from that time is probably an intensification of an activity that goes back at least another million years. The circumstances that made stone tools adaptive seem to coincide with a change to drier conditions; there was thus greater competition for food. The tools were used for both butchery and plant preparation, and show a clear understanding of the mechanical properties and geometry of the materials. In order to detach sharp-edged flakes, the core must be struck obliquely with some skill. No chimpanzee, not even Kanzi, can do anything like this. Making a tool requires a concept of cause and effect of which no animals other than humans are capable.

How could the earliest stone tool technologies have evolved, and what new brain processes were required? One possibility is that early humans, using stones to crack nuts in a similar manner to chimpanzees, could have smashed the stone by mistake and been impressed by the sharpness of such fragments. Perhaps they cut their hands by mistake. This could have opened the possibility of using the flakes themselves as tools for cutting. Early humans were probably more involved in scavenging than in actual hunting. Perhaps they used stones first to chase away animals that had killed, for example, a wildebeest, and then used the stone tools to cut up the body. Butchery is an important skill. There are suggestions that early stone tools, dating from about 2.5 million years ago, may have been used not only as choppers and scrapers, but also as objects to be thrown for the kill.

The first stone tools were essentially flaked and smashed-up quartz pebbles. The first known stone tool industry consisted of simple stone flakes, found initially at Olduvai Gorge in Tanzania. A more sophisticated industry, known as the Acheulian industry, was to develop later, dating from about 1.5 million years ago in Africa. The skills required for tool making probably led to handedness – the preferential use of one hand, which increases the possibility of skilled motor control. Acheulian tools include large cutting tools, picks, cleavers and bifacial hand axes, and there is also some evidence that the hand axes may have been hafted, that is, joined to a stick. A hand axe is made by removing flakes from either side of a stone. While these early tools were probably mainly used in relation to eating meat, there is evidence that they were also used for cutting and scraping wood, and for cutting reeds or grasses.

A key feature of hominid tools is the use of secondary tools, that is, objects used as tools in order to make another tool. Thus even simple stone tools require a hammer stone. It may be that stone hammers, like those used by chimpanzees to break hard-shelled fruit or nuts, were then used to shape rocks for cutting tools. Our tool-making ancestors had also to be competent field geologists in order to recognise which rocks were suitable for tool making. Some 2 million years ago, humans had acquired the not inconsiderable skill needed to make stone tools, which even for a modern human requires several hours to master. A carefully controlled sharp glancing blow is required to initiate a fracture in making the tool.

Progress in tool making was very slow, and may have required further evolution of the brain to give improved causal understanding. It took at least a million years to go from stone axes to other and more complex tools. This is a key point to which I shall return. The oldest wood tools are well-made javelin-like spears found in Germany and dating back some 400,000 years. Then by some 300,000 years ago, tool-making skills had accelerated, and by the Middle Stone Age there is clear evidence of hafted tools: composites with the components joined together. This was a major advance and I wish to argue that one clearly cannot make such a tool – joining quite different pieces together – without having a very clear concept of cause and effect. One would have to understand that the two pieces serve different purposes, and imagine how the tool could be used. One could not discover such a composite tool by chance. It was a fundamental advance in the technological revolution that actually makes us human, and at that time drove human evolution.

This technology, unlike making the simpler stone tools by flaking, which is essentially repetitive in nature, is complex and non-repetitive, and requires fine hand motor control to fit the components together. Human tool use was almost certainly socially cooperative. Sharing of new techniques would greatly advance the evolution of technology. Tool-using mothers for example, may have used them for food finding and preparation, and shared the process with their infants. Tools were also used for digging, to get at underground food such as roots, tubers and corms, and would also have been essential for working the skins of animals. These would be used for clothing, blankets, water containers and carrying devices. The earliest evidence for such uses dates back about 300,000 years. But agriculture seems only to have started some 10,000 years ago.

As already noted, progress with this technology was slow, and slivers of flint used as blades date back only 100,000 years. Bone artefacts for harpoons are of a similar age. Over the last 40,000 years, bone, antler and ivory were fashioned as tools, particularly for making pointed tools such as spears and harpoons. About 20,000 years ago, bows and arrows make their appearance, together with needles and sewing. The tools of the Upper Palaeolithic suggest a possible division of labour for tools that were more difficult to make. Blades, as distinct from flakes, were now common, as were composite tools, and the use of a variety of materials like bone. Such techniques took time to master. Moreover, some of the tools were no longer merely extensions of common bodily movements. A hammer is essentially a weighted fist. By contrast, using a saw involves recognising a quite new principle, and this is again true of the needle, which was of such importance to people surviving the last ice age.

The use of fire was a major 'invention'. Fire, how to spark it, presents a severe problem. Just how long ago humans began using fire is not clear: estimates vary from 300,000 to 1.5 million years. Use of fire requires clear causal beliefs about how to ignite a fire and keep it going. The ability to ignite a fire may have evolved over many generations. Striking two flints against each other, or rubbing wood together to give a spark, would most likely have occurred by chance. It was the recognition that this would ignite a fire that was crucial. Before the discovery of ignition, fire had to be borrowed from natural sources. This probably involved significant organisation of those who brought the fire and those who kept it going. This might have been one of the origins of market exchange, and might have led to the advantage of humans knowing about numbers – they could then bargain and trade in a reliable and fair manner.

Abundant artefacts are associated with the fully modern skeletons of some 40,000 years ago. They had not only stone tools but tools of bone, which were suitable for shaping into, for example, fishhooks. They had mastered the invention of sewing, and making rope and baskets. The tools used by these ancestors are rather complex. Anthropologists have tried to make these tools, and found that it required a five-step sequence as well as the initial selection of the right materials. Making such tools could not be done by imitation alone: instruction was most likely involved.

Our ancestors were consuming large quantities of meat and the stone tools were used to cut up carcasses. Planning ahead was essential, and our ancestors also needed an understanding of the environment they lived in, especially of animals and plants. They thus most likely used anthropomorphic thinking to predict how animals would behave, which is also true of modern hunters. Fruit collection is easily learned, extraction skills require more time to develop, and hunting is the most difficult foraging behaviour. It is clear that human hunting differs qualitatively from hunting by other animals. Unlike most animals, which either sit and wait to ambush prey, or use stealth and pursuit techniques, human hunters use a wealth of information to make context-specific decisions, both during the search phase of hunting and then after prey is encountered.

The modern human brain must have evolved in relation to the skill involved in tool use. That very slow improvement in tool-making abilities, over a million years, may have been related to the necessity for further adaptive changes in the brain. The earliest primitive human appeared about 1.8 million years ago with a brain size of 600 cc – 200 cc more than an ape. Homo sapiens, modern humans like us, only emerged 100,000 years ago with a brain that had increased from about 600 cc to 1,300 cc. The brains of modern humans are three times larger than those of other primates, and 75 per cent of the growth occurs after birth, while the corresponding figure in the chimpanzee is 40 per cent. Increase in brain size has costs. Neural tissue of the brain has the greatest nutritional requirement in the body, and so increase in brain size is a major metabolic burden, which could require new and improved technology to acquire food.

Both manual and vocal skills depend on programmed sequences in the brain. Areas of the frontal lobes linked with association and motor control, like the premotor area, have increased in size in humans compared to other primates.

Frontal lobe damage can lead to compulsive grasping or the inability to use a tool even while understanding its function. Chimpanzees do have basic motor skills and can be taught to trace complex patterns. Aimed throws are rare among monkeys, though chimpanzees and capuchins can do it.

But what about the hand itself? Sir Charles Bell in 1833 published his treatise on *The Hand, Its Mechanism and Vital Endorsements, as Evincing Design*. He used it to support his religious faith, for who else but God could have designed so amazing a machine? For him, the hand was central to human life. And E. O. Wilson, more than 150 years later, again argues that any theory of intelligence – and this must include belief, though he does not say so – must put at the fore the interdependence of brain and hand function. The human hand differs from apes in that it has a longer thumb and less curved finger bones. It is capable of both a power grip and a precision one: it can be used to wield a club or thread a needle. The early hominid hand had a short thumb, and the gorilla has one very similar to the human, but rarely uses tools. Apes do have a thumb that can be brought into contact with the index finger, but have great difficulty bringing it right across the hand, which we humans do with ease each time we grasp the handle of a hammer. Of course, freeing the hands from walking with the evolution of bipedalism was a crucial step.

It is also important to recognise that it is not just the shape of hands that matters, but also the ability of the brain to control their complex movements. Human manipulative skills are much greater than those of apes, and this is genetically determined because it is an intrinsic property of the brain. It has been suggested that the opposability of the thumb, and the associated wondrous dexterity, completely transformed our ancestors' relationship with external objects. This relationship could have promoted human consciousness, as the manipulation of objects became a self-conscious activity; once the individual becomes an agent operating on external objects in numerous different ways, causal beliefs are involved.

It is language that, in addition to causal beliefs, marks humans off from other species. Its role in human evolution, and its relationship to tool use and causal beliefs, is a central problem. Alarm calls by animals communicate, but contain no really factual information; and while bees do communicate information, they do not use a language. Unlike human language, animals communication systems are closed. People use language, not just to signal emotional states or territorial claims, but to shape each other's minds. Gestures may have been involved in the origin of language, but on their own they are not a language. Gorillas have been observed to make some thirty different gestures such as raising an arm before charging, putting their hands on those of another, sucking their lower lip and then backing away, and poking another's body. Most involve some response. Even so, the gestures never involve a third object, or point to objects distant from themselves. By contrast, their vocalisations are much less complex.

In what way could language be related to the making and use of tools? Most theories see language helping in the way tools are made and used. A few consider that it is tool use that actually led to language, or at least helped to lead to it. It is recognised that tools and language share some critical features: tool use is sequential – motor actions strung together – and in this it resembles speech, which consists of a sequence of utterances. It is formally analogous to grammatical language. Both speech and composite tool manufacture involve sequences of non-repetitive fine motor control. While making a composite tool is a bit like making a sentence, describing how to make one is almost a short story. From an evolutionary point of view, language could not arise out of nothing. It had to evolve out of neural structures or cognitive abilities already present; motor control is an obvious possibility. But another possible origin relates to tool use and to having causal beliefs. It helps in thinking about causality if one has language, but having language itself requires causal thinking, for that is what, for example, many verbs relate to.

Language has many metaphors, but two that are most commonly used relate to location in space, and cause and force. The equivalent of 'force' in causal belief may, in language, be a 'causal' or a 'doing' word, a verb. This relationship between verbs and causal thinking is an argument for believing that the evolution of language required causal thinking. Many verbs would not have any meaning unless the individuals using them had beliefs in causes and effects. Verbs ranging from 'go' to 'hit' to 'throw' require causal thinking. Thus, causal thinking preceded and was an essential prerequisite for language development. So we are back to the importance of tools as the driving force. Perhaps the slow progress in tool complexity is related to the evolution of language some 100,000 years ago. Language would help enormously with the construction and use of new tools, in terms of both cooperation and imaginative thinking.

Merlin Donald has argued that language emerged from the programming of the movements used to make tools and throw them. The actions used in making tools could have come to represent the tools themselves, and hunting and tool use could have been mimed. Note that children point before they speak. Note too, that signed language is very easily learned by children. Michael Corballis argues that human language arose from gesture rather than primate vocalisation. It is not a new view, and has its origins in the eighteenth century, when Darwin recognised its possible significance. Gestures could have been the forerunners of spoken language, but may have interfered with the tool-making process, thus giving vocal language a distinct advantage. Children simultaneously develop

hierarchical representations for both language and the manipulation of objects; they combine words into phrases at the same
age as they combine objects such as nuts and bolts. Moreover,
both are dependent on a region of the brain known as Broca's
area. Constructing objects is rather similar to language, and it
raises the question whether the construction of either involves
a similar set of beliefs.

Deaf children learning sign language can produce complex
responses, and this encourages one to believe that the origin of
language lies in gesture. What does the evolution of gesture
imply? Does it require the concept of cause and effect? The
growing recognition of signed languages as true languages, with
all of the expressiveness and generativity of spoken language,
provides a powerful boost to the idea that language originated
as a gestural system, and may even have evolved to a fully syn-
tactic system before being overtaken by speech. Deaf children
can create primitive language. In American Sign Language the
relation between signs and meanings are opaque: it is a code.
Deaf children acquire sign in the same graduated way and at a
similar stage to hearing children acquiring speech. And as the
great linguist Noam Chomsky has argued, sentences in every
language are built by the same rules, and children intuitively
know them.

About 100,000 years ago, tools had not advanced much,
though brain size had increased significantly. There may have
then been a protolanguage consisting of words, but no syntax.
The magic moment that may have led to new and more complex tools was possibly the arrival of syntax, and its origin must
lie in some special neuronal organisation. Thinking that
involved computations on internal representations of objects
became possible. Such imaginative thinking is essential for
complex tool making and use. In order to invent a spear with a
stone pointed tip, our ancestors had to be able to imagine its

construction, and what effect it might have. This adaptive advantage could be related to the start of organised hunting and a shift to a meat diet.

But does the ability to make complex tools in fact require language? In an experiment, students were divided into two groups and one was told verbally how to make a stone tool and given a practical demonstration, while the other group were taught by silent example alone. Both groups found it quite tough, but those taught in silence did just as well as the other group. Imitation may have initially been sufficient. Tool making and learning may not have been initially dependent on language. But it would have been difficult for early humans to acquire new beliefs from others without language. Trust in the reliability of such beliefs would have been an important issue, and could have helped establish the role of rulers and specialists.

Contrary to the emphasis I and others have given to tool use in human evolution, there is a quite widely held view that primate brain evolution has been driven principally by the demands of the social world rather than by the demands of interacting with the physical environment. The crucial difference between human cognition and that of other animals is the ability to collaborate and share goals and intentions. Robin Dunbar claims that there is a growing consensus that primate brain evolution has been driven principally by the demands of the social world and the environment and particularly by interactions with other members of the group. He argues that human brain growth, language, and intelligent behaviour were evolutionary changes related to the increasing social complexity of hominid community life. This argument is partly based on the increase in size of the neocortex, the part of the brain, that correlates with social skills such as mating behaviour, grooming and social play. Dunbar found that changes in the size of the neocortex correlate with changes in social variables, such as group size, rather than changes in the physical environment. There is also some evidence that neocortex size correlates with some measures of social skills such as mating behaviour. The neocortex is what may be regarded as that part of the brain involved in conscious thought. It occupies around 50 per cent of total brain volume in primates, and 80 per cent in humans.

Group size can be used as a measure of social complexity. Animals need to keep track of a constantly changing social world. Some increase in brain size may have been necessary to ensure appropriate social interactions as shown by mutual grooming, on which primates can spend several hours a day. For Dunbar, the evolution of language is intimately linked to its ability to facilitate the bonding of larger groups, and coopera-tion within them. But without causal thinking about the inter-actions of objects, I find it hard to see how improved social understanding could have been a real advantage, or how it could have led to technology.

'Man is a tool-using animal . . . Without tools he is nothing, with tools he is all,' wrote Thomas Carlyle (1795–1881) and in 1941 Julian Huxley pointed out that 'There is no essential dif-ference between man's conscious use of a chipped flint as an implement and his design of the most elaborate machine . . .' How right he is. The manufacture of stone tools some 2.5 mil-lion years ago was indeed an elaborate procedure, and then some 100,000 years ago tools became more complex, perhaps because of language. What is amazing is that a mere 20,000 years separate the first bow and arrow from the International Space Station. Belief in cause and effect has had the most enor-mous effect on human evolution, both physical and cultural. Tool use, with language, has transformed human evolution and led to what we now think of as beliefs.

CHAPTER SIX

Believing

> You believe that which you hope for earnestly.
> Terence – Roman playwright (BC 190-159)

The evolution of the skills for tool making and the use of tools, together with language, opened up a whole new set of mental operations. Humans were now thinking about the causes involved in all sorts of activities: hunting, food gathering, social relationships, illness, probably dreams, and even life and death itself. This, I am proposing, is the origin of what we now call beliefs. I must admit that the transition from understanding cause and effect in relation to tool use, to trying to understand the weather or death, is not easy to explain, and probably required creative thinking. It is possible that the evolution of consciousness and language could have been involved. It has been argued that people experience consciousness because they are aware of their own causal actions.

Causal beliefs are nevertheless fundamental, since a major feature of belief is that it is used to guide how we behave, and so is at the very core of our existence. One can think of it as an explanatory tool. When one refers to someone having a belief, we think we can quite reliably predict how that belief will determine their behaviour in particular circumstances. This view is technically called the intentional stance. This also implies that a person is aware of their own beliefs.

The anthropologist Clifford Geertz has made the important point that humans are incapable of looking at the world 'in dumb astonishment or blind apathy'. and they always try to explain what is going on. This is fundamental to all thinking about beliefs, particularly those relating to causes that affect our lives. Story-telling, particularly about causes, is very important for us; there are studies that found that about 15 per cent of our day-to-day conversations contain ideas, that is, beliefs, about causes. We take facts from everyday experience and weave a story about them, which is sometimes true, sometimes less so. This is our way of keeping ourselves believing that what we do is good, and that we are in control of our behaviour. This patterning and weaving of our beliefs is fundamental to our thinking, and will be considered in much greater detail when we come to religious beliefs, and those relating to the paranormal and health.

We have seen that causal beliefs are a developmental primi-tive – they develop early in life and are initially little related to culture, and are acquired with a minimum of effort. Here I will focus on some of the processes by which adults, mainly in West-ern cultures, acquire some of their everyday beliefs. In all cultures explanation of events of importance can be very pleasurable; it has even been suggested that explanation is to cognition what orgasm is to reproduction.

How we arrive at beliefs is far from clear; it is a mixture of experience, cognition, intuition and emotion. In constructing beliefs we use current knowledge as well as trying to retrieve relevant information; we also try to deal with consistency and inconsistency, both internal, and in relation to the views of others. These attempts are often unsuccessful, as many beliefs are founded on inadequate or insufficient evidence. The evi-dence required in science is of a quite different kind from that involved in common beliefs. The special nature of scientific beliefs will be considered later.

The processes by which we arrive at our beliefs are beset with logical problems that include overdependence on authority, overemphasis on coincidences, distortion of the evidence, circular reasoning, use of anecdotes, ignorance of science and failures in logic itself. One must always bear in mind Hume: 'That no testimony is sufficient to establish a miracle, unless the testimony be of such a kind, that its falsehood would be more miraculous than the fact which it endeavours to establish.'

Evidence is itself not easy to evaluate, and the many books and papers written on intelligence, the evaluation of evidence, and rationality, still leave it resistant to a simple definition. Explaining why a particular conclusion was drawn from a mass of evidence can be very difficult. The calculation of the probability of evidence being reliable is no trivial matter; statisticians spend their lives on such problems. These issues are largely beyond the scope of this book, and when I talk about reliable evidence I hope not to be too controversial, but my own beliefs will undoubtedly come through.

Beliefs, once acquired, have a kind of inertia in that there is a preference to alter them as little as possible. There is a tendency to reject evidence or ideas that are inconsistent with current beliefs, particularly if they undermine central beliefs; this is known as the principle of conservatism. Cognitive consistency is another central idea in the psychology of belief. We are seen as information processors who seek cognitive coherence. A satisfying internal consistency about our beliefs – the stories we tell ourselves – is a fundamental part of human nature. Cognitive dissonance refers to a universal cognitive process that makes us feel very ill at ease when we recognise that we hold conflicting beliefs, and so we try to reduce the conflict. We try, in our beliefs, to tell a consistent tale. Whenever there are two ideas that are psychologically inconsistent, there is a mechanism driving us to reduce this dissonance, and thus change ideas or beliefs. So a smoker who also believes that smoking causes cancer might convince him or herself that it is an essential activity. There is some evidence that the differences between the left and right hemispheres of our brains are involved in this process. The right hemisphere feeds information into the left, and it is the left that has to form beliefs and deal with cognitive dissonance. We may only change our mind when the information from the right overwhelms the beliefs held by the left brain. We nevertheless have the capacity for self-deception, which includes the ability to hold two contradictory beliefs at the same time.

One view of how we acquire beliefs is that we first collect the facts, which we check carefully, and then, using them and logical reasoning, draw inferences. Rational beliefs can come from such a process. However, it is more common not to proceed along these lines. Instead we tend to believe even when we do not know the relevant and necessary facts. Often even the foundations for the belief remain hidden. There are processes in our brain that allow us to reach conclusions without conscious rational thinking. On this basis, belief is the result of intuition, though there may also be some checking of the facts surrounding the belief.

Humans have a strong tendency to construct novel interpretations to cope with the limited, even false, information available to them. There is, alas, not so clear a dividing line between the beliefs of the sane and those with mental illness, which can be seen to be clearly false. There are indeed pathological false beliefs that need to be understood. For example, during a florid episode, a schizophrenic patient may retain a good memory of past events, but become very confused about the boundary between the self and another person. The patient may respond to a tactile stimulus with the belief that the sensation belongs to someone else, or that it exists outside the body. Voices and thoughts can also appear to come from someone else. Confabulation, a false belief, is not uncommon, and is used to make sense of false memories and delusion. I return to this in the next chapter.

There is evidence that simple learning may play a role in generating beliefs. Associative learning comes about when an event occurs that is associated with a quite different event. So, if some event, rain or plague, occurs close to some other special event – perhaps a rain dance or failure to follow a religious ritual – a causal connection may be believed to have been established. Regularity of an association between two events provides a strong basis for assuming a causal interaction. This is very different to actual causal explanation based on what one believes are the relevant forces for physical interactions, or for other humans, the beliefs and desires of the person whose behaviour one is trying to explain.

Beliefs are held in one's memory and can be recalled. We express beliefs even when, all too often, we do not have the evidence, knowledge, or facts to support them. Moreover, emotions can undoubtedly influence our beliefs. In addition, the distinction between knowledge and belief becomes less clear in relation to memory. It can be difficult to distinguish between belief about what happened, and the memory of what occurred. Beliefs can in fact be constructed from memories, and mere repetition of a statement can strengthen one's belief that it is true. Beliefs can be very hard to change. There is also evidence for social conformity in beliefs and of the importance of authority in determining these beliefs.

A major contribution to our understanding of everyday judgements, and thus beliefs, has been to examine what are called heuristics and biases. Heuristics is the technical term used for the simple rules humans use for making fast decisions in their day-to-day lives: they are essentially simple beliefs. The core idea of the heuristics and biases programme initiated by Kahneman and Tversky is that judgement, and the development of beliefs with uncertain evidence, is usually based on a small number of heuristics and cognitive procedures. They have identified three general-purpose heuristics: availability, representativeness and anchoring. Judgements based on these are made, it seems, on a mixture of intuition and reasoning. It is also a quite general feature that there is overconfidence in the correctness of our judgements. There is, they claim, a tendency built into our cognitive machinery to find order in ambiguous stimuli – we seek patterns even where there may be none. But finding patterns in nature was a great evolutionary advantage and is part of our understanding of cause and effect.

Availability is related to the ease with which the information comes to mind and is accessible. The true likelihood of an event can be thought of in terms of probability theory, and is based on knowing the total possible number of events and so being able to determine the chance of a single event taking place. People do not naturally think like this: we tend to rely more on the information that most easily comes to mind – availability is a dominant feature. This is particularly true when people make estimates of risk, as already discussed. When a hypothetical outcome is imagined or talked about, the individual then believes it is more likely to occur. In one study, subjects were asked to imagine being arrested for a crime, which made them believe more strongly that it would happen to them.

This is explained by the availability heuristic: cognitive availability. Evidence for this comes from a study in which students were asked to imagine themselves getting a disease called Hyposceried-B in the future. Some were told that the disease has common symptoms like headaches and muscle pain. With another group, the symptoms were less concrete: an inflamed liver and a vague sense of disorientation. When questioned later, the second group were much less likely than the first group to believe that they could contract the illness.

Another key heuristic is representativeness, by which we judge something new in terms of how well it matches what we already know. It provides a means for finding pattern in, for example, random sequences of events. This is the process by which we tend to assess the similarity of outcomes and events on a quite superficial basis. There is an assumption that like goes with like, so causes will be like the result – heartburn will be caused by spicy food. Representativeness probably plays a role in generating the Barnum effect. There are also, as we shall see, beliefs in the curative powers of certain medicines because they seem to bear some relationship to the illness. Again, consider the belief that 'You are what you eat.' In Papua New Guinea the locals believe that the faster the food grows, the faster you will put on weight if you eat it; and a quite widely held belief by students is that individuals tend to acquire the characteristics of the food they eat. A further aspect of representativeness is that it might have an evolutionary advantage, as our ancestors preferred the familiar – which they knew to be safe – to the unfamiliar, and so it may also be wise to have beliefs similar to each other.

Judgement of a person's character can, because of representativeness, be greatly influenced by some key phrases describing the person. When given a description of a beautiful young woman, it would not enter anyone's head that she is an engineer. Again, when a man was described as a thirty-four-year-old who was good at maths, lifeless and unimaginative, few thought he would play jazz for a hobby. Most of us, when asked if there are more words with k as the first letter, than with k as the third, will go for the former; but actually there are three times as many with k as the third letter. The evolutionary origin of representativeness being programmed into our minds might be that it was adaptive in a world where rapid decisions needed to be made; the most reliable interpretation of new events was to relate them to what was already known.

Representativeness influences judgement of how much a sample corresponds with the population as a whole. With representativeness, individual confirming examples are given undue weight in supporting a belief. A nice example is provided by the following question given to university students:

> A health survey was conducted on a group of males in Canada, of all ages and occupations. Please give your best estimate of the following findings:
>
> What percentage of the men surveyed have had one or more heart attacks? What percentage of the men surveyed are both over 55 years old and have had one or more heart attacks?

The mean estimates given were 18 per cent and 30 per cent respectively, which is, of course, impossible, as there cannot be more men older than 55 with heart attacks than all men who have the same illness. Kahneman and Tversky do not accept the suggestion that inside every muddled and incoherent person is a very rational one trying to get out.

Representativeness may be one of the reasons why science, as will be discussed, can go against common sense and so appear almost unnatural. In science, like may not go with like. When, for example, it was first suggested that malaria was carried by a mosquito, it was not accepted, probably because it went against representativeness that a tiny insect could cause such a terrible illness.

The third important heuristic affecting decision and beliefs is anchoring. This refers to the process in which an uninformative number has an effect. For example, a wheel of fortune is spun and then stops at, say, 65; the subjects are then asked if the percentage of African countries in the United Nations is above or below that number. Remarkably, groups for whom the wheel stopped at larger numbers gave higher estimates than those for whom it stopped at lower numbers. A somewhat more understandable example is asking subjects for a rapid estimate, with no time to actually do the multiplication, of 8 x 7 x 6 x 5 x 4 x 3 x 2 x 1, or 1 x 2 x 3 x 4 x 5 x 6 x 7 x 8. The former estimate is almost always higher; and both are far too low, as the answer is 40,320!

In general, when making decisions, people rarely test their belief on evidence that might show it to be wrong. We all seek confirmation of our beliefs rather than trying to falsify them, as Terence, quoted at the beginning of this chapter, so clearly understood. The anthropologist George Frazer, author of *The Golden Bough*, explained the practices of magic as an attempt to relate cause and effect, and argued that there was, as is so common, a focus on successes, and that failures were discounted. In Peter Wason's classic experiment, subjects are asked which card to turn over to test the rule that if there is a vowel on one side, there is an even number on the other. The four cards are Ace, King, 2 and 7. Most turn over the ace, just trying to confirm the rule. But then they can try one more card, and most turn over the 2, which tells us nothing; only 4 per cent turn over the 7, which could falsify the rule. Negative evidence is too often neglected, and a lovely counter-example comes from the Conan Doyle story about the theft of a racehorse. Sherlock Holmes noted the significance of the fact that the dog had not barked. In the real world it is rare that a single counter-example to a belief can be that crucial.

Emotions play an important role in relation to belief. For example, a person may feel an emotion like jealousy because of a false belief, and the emotion should not persist once the belief is known to be false. But the emotion can, in fact, strengthen the false belief. Emotions like jealousy or envy can even generate false beliefs. Intellectual conviction is thus just one driving force for beliefs. Many beliefs are acquired during childhood from parents, friends, and other cultural influences. Beliefs can be acquired by accepting on trust the beliefs and words of somebody one regards as an authority, and by learning from personal experience. For some people, beliefs are like possessions, and can evoke the feeling of ownership and attachment that one normally feels for material possessions. 'I used to be religious, but gave it up because it did not help.' Are there unconscious beliefs as distinct from conscious ones, and could these affect our behaviour as well as our conscious beliefs? This is of course at the heart of much of psychoanalysis; we have, the analysts claim, beliefs that we do not acknowledge, but an analyst can bring them into consciousness. Evidence for this is hard to find.

Schumaker draws a distinction between primary reality, the world as it really is, and personal reality, which biases the individual's perception of the primary reality. The result is that much of the information in an individual's mind is false, and is strongly influenced by the local culture and beliefs. Going further, he argues that this could even be healthy. This is quite contrary to the view that a mentally healthy person is one who sees the world as it actually is. Schumaker's view is that some self-deception correlates with mental good health – lying to oneself may be adaptive. Humans can benefit from misreading reality under certain circumstances.

How important is rationality and logic in our construction and acceptance of beliefs? Is it necessary to invoke the concept of rationality to explain or understand or justify or falsify people's beliefs? By definition, a belief is a proposition to which a truth-value can be attached. In all languages it is possible to either deny or assert the truth of some or other proposition. But there are some beliefs that are so weird that they are never put forward – that a pin in a matchbox is identical with the

Empire State Building, for example. But is it not possible to believe that one could be changed into the other? I was recently having a discussion with a theoretical physicist about evolution, and whether it would ever be possible for humans to evolve a pair of wings, given as much time as necessary. He thought it was, given enough time. I pointed out that it was so improbable as not to be taken seriously, but he persisted that while the probability was very very small, it was nevertheless finite, and all changes were possible. I suggested that his argument required him to believe that the next day he could wake up as a dog. Yes, he agreed, again very very very unlikely, but in terms of physics, possible.

What about logic and contradiction? When in 1903 Levy Bruhl read the translation of a Chinese historical work, he was so puzzled by it that he wondered if the rules of Chinese logic were the same as those in the West that had their origin in ancient Greece. Was Chinese logic really different? The main conclusions from his investigations into cultural differences, particularly among the indigenous societies of America and Africa, was that there was no evidence of a universal logic dominating thinking. Participation in group beliefs was much more important, however incomprehensible these might be. The mentality was often pre-logical, in that contradiction did not matter. Yet normal 'logical' thinking was pervasive in everyday activities.

Across cultures, there are two relatively distinct classes of causal explanation. One is based essentially on the properties of the object, so some bodies float because they are light, or a person's behaviour is something embedded in their nature. The other class of explanation assigns external forces, like leaves being blown by the wind, and holds that personal behaviour is determined by social forces. In Zatopec culture, a person's behaviour is explained in terms of the social situation, as in Hindu culture. Americans consider that the person's disposition as largely determines their behaviour. An interesting example of different cultural beliefs is provided by the analysis of two accounts of why two murders took place, one in the *New York Times*, the other in a Chinese language newspaper, the *World Journal*. The *Times* report attributed a major cause to the personal disposition of the murderer, namely that he was mentally unbalanced, while the Chinese paper emphasised the social situation, the selfishness and violence of American society. Such differences are much smaller in relation to beliefs in the physical domain, such as the movement of objects.

There is quite often a refusal to believe something even though the evidence is overwhelming, like the fact that one's partner has been unfaithful, or that one's child has become a drug addict, or that one has an illness that is fatal. This raises the question of the extent to which we are free to choose what we believe. The developing child's beliefs in physical causes and effects are not conscious choices. Just how voluntary our beliefs are is far from clear. Things are made difficult if Hume is correct when he says that we cannot choose what to believe, since it comes from causes affecting our thinking over which we have no direct control. Can we really not choose what to believe? Could a God believer choose one day to be an atheist? And then there is the further problem of self-deception, and the way people cultivate a belief by seeking evidence for it. William James (1890) wrote: 'whilst part of what we perceive comes through our senses from the object before us, another part always comes out of our mind.' Memory thus plays a key role in determining the action of the mind, since it provides prior knowledge.

Particularly difficult is the origin of a fundamental belief in the theory of mind; that is, being able to have beliefs about the beliefs of other people, and thus how they will act. Individuals who are autistic do not have this ability. This condition is largely genetically determined and there is a relatively high incidence of autistic children, about 6 per 1,000. But knowing what another person believes can be quite difficult. They may be lying, or unable to express their belief. And if the stated belief is inconsistent with their behaviour, there is a problem: what if someone believes that acupuncture is of no value but goes to an acupuncturist when ill? If someone kills in self-defence, did they really believe their own life was in danger? This is highly relevant, as the reason for our interest in the beliefs of others is largely to explain and predict how they will behave.

Since belief depends on the activity of the nerves in the brain, then reliable though evolution and biology may be, things can go wrong, as in autism. We will now consider other examples where abnormalities in brain function can lead to what are incontestably false beliefs.

CHAPTER SEVEN

False

> I discovered that beauty, revelation, sensuality, the cellular history of
> the past, God, the Devil – all lie inside my body, outside my mind.
>
> Timothy Leary

When the brain's belief system clearly goes wrong it can throw light on the normal process. Certain beliefs are so palpably false that they are clearly abnormal, and must have a biological basis, primarily due to brain malfunction. Studying these could thus be very helpful to our understanding of the normal processes in the brain by which beliefs are acquired. Evolution has selected humans to have brain mechanisms that can give rise to reliable causal beliefs, particularly in relation to manipulating the physical world. But these mechanisms are complex and can go wrong, particularly in relation to memory and other beliefs. This will be illustrated with examples of neurological abnormalities, psychiatric illnesses, psychedelic drugs and hypnosis.

False beliefs arising from brain pathology most often involve beliefs about the subjects themselves or other people, and less often involve certain aspects of the physical world – no cases are reported where someone believes they can drive a nail into wood with a feather, or put the Eiffel Tower in their pocket. The nature of false beliefs as a result of brain abnormalities suggests that some are related to pre-existing programmes in the brain, which are easily activated, so we need to understand the normal functions of these programmes, and their evolutionary significance. False beliefs illustrate how damage to the brain can result in irrational beliefs, and one can imagine that in normal life a minor abnormality in function could have a similar effect. The processes that seem to me to be most helpful to our understanding of beliefs are confabulation, delusion, hypnosis and schizophrenia, and the effect of certain drugs. In almost every case the individual will try to give a causal explanation for these beliefs.

Of particular importance is confabulation; that is, finding explanations for our experiences and conditions that have little relation to what has actually happened. It is probably closer to the way we normally think than we may like to believe. Confabulation is the phenomenon whereby patients may produce false memories. For example, the patient may relate in graphic detail how his or her parents visited last night, and later it becomes clear that the mother died four years ago, and the father twenty years ago! Patients also recall incidents after hearing a story, though they were not part of the story. Such people are not aware that they are producing false memories. These memory distortions can range from the benign, being sure you posted the cheque, to a serious, confusing what you actually saw at the scene of the crime. There is also the fantastic, like recalling abduction by aliens.

We humans want a consistent story about events in our lives, a plausible explanation, and all too often invent one to fit with other beliefs. Our personal memories are very dear to us and we often have a strong belief that particular memories are true, reliable. When two people disagree about a shared past event, each is often quite sure the other is wrong. We can confabulate in order to make sense of the past and present. With brain damage, distortions of memory are far beyond the normal range and can even be bizarre. Another patient, for example, told false stories of how members of his family had been killed before his eyes. There are probably differing kinds of mechanisms that underlie these forms of confabulation.

Spontaneous confabulation, in which weird and false ideas may rove over a number of themes, seems to be the result of damage to the frontal lobes of the brain, particularly in the bottom region. It is much more common when the patient's memory is very obviously poor. Confabulations might result from an inadequate ability to evaluate memories properly, failure to use common and general knowledge of the world, or failure to recall negative evidence. A man with brain damage is in hospital, and suffers from confabulation. When asked by the doctor where he thinks he is, he replies he is at work. But, says the doctor, who are all these people in the ward? My employees, is the reply. But they are in bed. Yes, he says, we like them to be comfortable. One patient believed he was a Russian chess master, though he could neither play chess nor speak Russian. His explanation was that he had been hypnotised to forget that he could speak Russian.

A distinction is drawn between momentary confabulation, which consists of real memories but not in their correct temporal order, and fantastic confabulations, which are bizarre and bear little relationship to real events. In both cases there is an attempt at internal consistency, and most often the topic is autobiographical. Thus, it is the normal memory retrieval processes that have become disordered. Even in normal subjects, long-term memory of key events can be faulty.

Could confabulation be a more common part of our normal belief systems than is generally recognised? That is, we construct apparently coherent stories about what happened, which are at the edge of being false, but where consistency and internal satisfaction have to compete with testing against the real world, we choose consistency. We all too often make such errors in our daily lives, but in patients with brain damage this happens in a much more dramatic and recognisable manner. Perhaps we are all, to some degree, confabulators.

Among the false beliefs that can arise from mental disorders are delusions and hallucinations. A delusion is defined as a belief that is firmly held on inadequate grounds and is not affected by rational argument or evidence to the contrary. Common religious beliefs are excluded from this diagnosis, probably as they are considered to be largely culturally determined, and not the peculiar or special belief of an individual, and they are considered in detail in the next chapter. The relevance to our beliefs is that delusions and hallucinations can be due to the brain generating such experiences when damaged or functioning abnormally. The implication is that such weird experiences are just modifications of normal brain function, and that less extreme examples than those that occur in patients could easily occur in daily life, and thus affect our beliefs.

There are a number of neurological illnesses that result in delusions involving false beliefs similar to confabulation, though memory is not at the core. Patients who have had a stroke that has affected the right hemisphere of their brain, so that their left side is paralysed, may deny on occasion that they are paralysed. For example, an elderly woman, who can neither walk nor use her left hand, will say that she can do both. When asked to clap her hands, she will make the movement with her right hand and say that she is indeed clapping. This is similar to confabulation and is typical. Other patients, when asked to point with their left hand, say they cannot because of arthritic pain, or 'I've never been very ambidextrous.' As the neurologist Ramachandran says, to listen to a patient deny ownership of her own arm and at the same time to admit it is attached to her shoulder is, for the neurologist, perplexing in the extreme. Patients may recover from the brain damage and then stop denying they are paralysed. When questioned as to why they had a false belief, some deny that they had such a belief, whereas others may say that their mind knew it to be false, but would not accept it.

The Capgras delusion is another example of a neurological condition giving rise to false beliefs. When the patient sees someone he knows very well, a wife or parent, or child, he claims that the person looks like, for example, his spouse, but she is not really his wife and may be an alien impostor. In other respects the patient may be largely normal. One explanation is that in seeing his wife, he recognises her, but that the normal emotional response is absent and thus it could not be his wife, and so he believes it must be an alien pretending to be her. More generally, this reflects a dissociation between recognition and familiarity. A related but different disorder is prostopagnosia, in which the patient cannot recognise the identity of faces. Yet physiological studies show that the patient does respond to a familiar face, even though he or she fails to recognise the face as someone well known to them. The Fregoli delusion is one in which patients keep seeing someone close to them, possibly a family member, who follows them around but is unrecognisable because they are in disguise.

There are no sharp dividing lines between normal beliefs and delusional beliefs. There are symptoms of delusions in the general population, even in individuals who appear to be perfectly normal. Several studies have found that 10–15 per cent of the population have hallucinatory experiences in their lives, and 20 per cent reported delusions. A study of delusions in the general population made use of a delusions inventory. Questions in the delusions inventory included asking whether the subject has the feeling that they can read other people's minds, that other people can read theirs, that someone is aiming to harm them, that there is some mysterious power working for the good of the world, and that they are not in control of their own actions. For each of the forty questions, the subject was asked to indicate the strength of the associated belief. The range of scores overlapped considerably between normal and clinically deluded subjects. About 10 per cent of the normal group had scores above the average of the deluded group. The boundary between those with normal and those with abnormal beliefs is fuzzy. All these studies imply that humans have a natural tendency to somewhat mystical explanations, which are normally constrained, but can emerge under certain circumstances.

Some delusions are thought to arise in the following way. First, due possibly to brain damage, the individual has a very unusual experience, which may be a hallucination, or even an episode of out-of-body feeling. Then a hypothesis is generated about the cause of this experience, and accepting the reality of the experience, the individual forms a belief about the cause, which may have little to do with normal experience. This is a crucial stage that is not well understood, but once formed, the belief is resistant to change. In judging a task involving simple probability judgement, subjects who scored higher on the delusion inventory showed what is known as the jump-to-response. This refers to reaching conclusions on minimal evidence. They also showed a need for closure, that is, the determination to give an answer rather than remain uncertain or experience ambiguity. There are similarities to the way we normally construct beliefs. A very different approach to delusions comes from psychoanalysis, and suggests that delusions are defence mechanisms to deal with unpalatable truths; for example, the Capgras delusion is due to hidden hate of the partner. But this, I believe, is little more than psychobabble.

There are many examples of delusions that range from pathological jealousy to the belief that one is changing sex. Raj Persaud, a psychiatrist, has recently described many of the delusions found in patients with mental illness, and suggests they are more common than is generally supposed. There is even an argument that they can be adaptive, by providing a more positive view of the world, for example, being blind to one's partner's faults and thereby maintaining a marriage. Thus, delusions could be seen as an attempt, alas a false one, to make sense of the world; another example perhaps, of confabulation. A feature of delusions is that they can lead to behaviour that prevents their falsification. For example, those with persecutory delusions repeatedly took unnecessary precautions, which then convinced them of their value. The list of delusions is long, and some examples of case histories described in Persaud's book include a description of a man who believed he was a crazy tiger, and a woman who believed she had made love to the devil.

Patients with schizophrenia, acute mania and psychotic depression can have bizarre delusions, including religious ones. For example, there may be the belief that they are reincarnated as Jesus, or some other divine power; there may be voices from God or the devil giving instructions on what to do; there may be the belief that an unpardonable sin has been committed. They may hear a voice telling them that they cannot do what they want to do, or telling them to kill God. Sometimes they think another person is speaking for them, or that they are victimised and someone is trying to hypnotise and kill them. Schizophrenic delusions may result from the patient experiencing a thought but being unaware of where it came from: it is alien, and so requires a causal explanation. It would be helpful to know how these compare with other delusions: are they special, and how should they be understood? False beliefs are clearly held by schizophrenic patients. About 1 per cent of the population will have such an episode, and there is a very strong genetic basis for this.

There are, it is claimed, certain common characteristics of mysticism in all cultures and religions throughout history. These are a feeling of blessedness and peace; a sense of the holy, sacred or divine; and a sense of the paradox of life, together with a sense of the ineffable. Mystical experiences are, for the individual, very real, as real as those of people who suffer from mental illness and have false beliefs. Moreover, evidence that human brains are prone to hallucinations, similar to those experienced in schizophrenia, comes from studies which found that about 40 per cent of American college students reported having heard their thoughts spoken out loud, and 5 per cent had actually held conversations with those voices. Only about 30 per cent had no experience at all of hallucinations.

One should try to distinguish between a false perception and a false belief, even though they may be closely linked. A patient who believes that his thoughts are being spoken is suffering from a false perception, while one who claims his colleagues are poisoning him has a false belief. Often it is difficult to distinguish between false beliefs and false perceptions, but as William James long ago recognised, part of what we perceive comes through our senses, while beliefs are constructed in our mind. Can abnormal perceptions explain abnormal beliefs? Not easily, as bizarre beliefs like Capgras, in which a spouse is viewed as an alien, are stubbornly held, even though the perception is normal. Experiments with voice distortion show that normal people come up with simple explanations, but patients in the acute phase of schizophrenia talk of evil spirits, and other bizarre rationales; they may have problems with semantic memory and are less concerned with meaningfulness.

Hallucinations may be visual or auditory. They can occur in healthy people, usually when falling asleep. Some hallucinations can result in the experience of seeing one's own body projected into space, even just in front of oneself – an out-of-body experience. Drugs or brain damage can result in hallucination. One patient who had had a stroke saw coloured letters and numbers superimposing themselves on objects in his view. Later, he had the impression that all the other patients had plastic tubing connecting their mouths to their ears. He then began to see seagulls everywhere. Visual hallucinations are common following the onset of blindness in later life.

Delusions are common in patients with schizophrenia. Paranoia is not uncommon, and provides another example of false beliefs in which the individual has a very distorted view of the real world. There is also evidence for loss of neurones in specific brain regions, and this may result in the inability of the brain to properly sort out incoming information. The patients may thus experience the world as overwhelming, and commonly believe that an evil force is controlling them. There is good evidence that schizophrenia has a strong genetic component, and that increased transmission by the neurotransmitter dopamine is associated with the psychotic state. Drugs that increase this transmission make things worse, while treatment with drugs that reduce it can be quite successful.

Just a change in the concentration of this transmitter can thus have profound effects on beliefs, and this again shows that the underlying circuits are crucial and have some adaptive value when working normally. This suggests that such belief systems are present in our brains, but normally inhibited, for how else could just changing the concentration of a chemical have such profound effects? One possibility is that these circuits are involved in the ability to predict the behaviour of other people, and only when acting improperly lead to the individual finding secret meanings in other people's most casual gestures, and having grand theories about external influences. False beliefs can result from the improper function of normal and adaptive belief mechanisms.

Delusions of control are only one class of symptoms in schizophrenia – patients feels that their own actions are being created not by themselves but by some outside force, or that emotions are being manipulated by outside forces. The delusion of being controlled by an outside agent could be due to the uncoupling of the intention to move from that action itself. Some patients have been shown to be unable to monitor their own movements without visual cues. This fits with the theory of motor control in which, in order to monitor our actions, it is necessary to monitor the sensory consequences of those actions. The programme for generating movement also generates the predicted sensory consequences, but if something goes wrong there could be a mismatch. This could lead to a patient being unaware of motor disabilities, as described for the paralysed patient earlier. Brain imaging has shown that identical active movements are processed differently in the brain, depending on whether they are attributed to the self, or to an external source. Overactivation of the cerebellar-parietal network can lead to misattribution of the movement to an external source.

Severe depression provides a further example of pathological false beliefs, and can provide a nice example of how it is related to a normal function: the emotion sadness. The psychoanalyst Aaron Beck realised that it was the conscious thoughts of his depressed patients that really mattered. Instead of the psycho-analytical assumption that it is the unconscious thoughts that maintain the depression, Beck recognised the fundamental importance of automatic negative thinking by his patients. All their beliefs are negative and may have little relation to reality. Negative thoughts can permeate a patient's mind and can result in false beliefs. These negative core beliefs are usually global, over-general, and absolute – there is no doubt.

Depression, too, has a strong genetic basis. Many episodes of depression are triggered by loss, which results in sadness, a normal emotion that leads to attempts to make up the loss. Depression can be viewed as sadness becoming excessive, even malignant, and so giving rise to false beliefs. The cognitive processes in the brain interpret the emotion in this false manner and feed back to make the individual even sadder, and too often this can lead to suicide.

In the inner world of the depressive, the self is perceived to be ineffective and inadequate, whereas the outside world is seen as presenting insuperable obstacles; moreover, there is the belief that the depression will continue forever, and that the patient will never get better. The sufferers draw negative conclusions without any evidence to support them – 'I failed once, and this means I will never be successful'; and reach important conclusions on the basis of a single event – 'John says he does not love me, nobody cares for me.' Underlying these negative thoughts are a set of false beliefs, and it is the aim of cognitive therapy for depression to uncover and correct these beliefs.

The other side of the coin of false beliefs in depression is mania. Mania results in overactivity and feelings of great elation and self-importance, often with false beliefs. Mania has even been described as having a mystical quality, an example of which is that described by the writer Theodore Roethke. One day he felt good, and then felt that he knew what it was like to be a rabbit, and then a lion; so he entered a restaurant and ordered and ate raw meat. Kay Redfield Jameson bought twenty Penguin books in order to form a colony of penguins, and the poet Robert Lowell believed on one occasion that he might be the reincarnation of the Holy Ghost and could, if he wished, paralyse cars. John Ruskin had the experience of light and sound becoming harmonious, as well as very intense and sometimes frightening experiences.

General anxiety disorder provides another example of false beliefs related to a normal cognitive process: being aware of danger. It is not focused on a particular issue, and the individual can believe a variety of situations to be threatening. In anxiety disorder there is a fixation on the concept of danger and impending loss of some sort, together with the belief of a personal inability to cope with the danger. The symptoms of anxiety may themselves be interpreted as a sign of a serious illness. In panic disorder, patients interpret these bodily sensations in a very negative manner. Again, a person anxious about their health resorts to numerous medical consultations, and so increases the likelihood of receiving negative or misleading information. All these are cognitive errors or false beliefs and involve, for example, drawing a conclusion when there is very little evidence, focusing on just one aspect of a situation, personalising external events, and giving negative attributions to other people's behaviour.

Obsessive–compulsive patients have a compulsive belief that they must carry out some action or think some thought, which at the same time they feel must be resisted. They may hold beliefs about the unacceptability of certain types of thought, and believe that certain thoughts can lead to disaster, or that disaster can be avoided by some sort of magical ritual. Common obsessions are about contamination, which can lead to ritual hand washing. There are also aggressive impulses, and recurrent sexual imaging. Such delusions are beliefs held with complete conviction, even in the presence of contradictory evidence. One woman was obsessed with the belief that her husband would swallow a fragment of glass and so bleed to death. She compulsively examined all dishes and glasses in the house, even the carpets, for such a fragment.

The eating disorder anorexia nervosa involves a false belief about one's body. Anorexics believe themselves to be too fat, even when severely underweight. The anorexic – usually an adolescent girl – shrinks her world by narrowing her perceptions to thinness. She may do this to avoid entering the adult world, or as a means of giving herself the feeling of control. Anorexics exhibit just the opposite to suggestibility – they are totally non-suggestive and accept nothing told to them relating to their condition. They have a closed system of thought of an autohypnotic nature. Anorexia is a condition absent in non-Western cultures. However, there is in Asia a mental disturbance known as koro, which affects men who believe that their penis is shrinking, will enter their abdomen, and that they will then die. Schumaker goes so far as to suggest that anorexia nervosa is an excellent model for trying to understand paranormal beliefs, to which I shall return in a later chapter.

Psychedelic drugs can radically affect visual perception and produce hallucinations as well as delusions. This again means that there are neural circuits that can be activated to produce such effects, and that they could in principle become active without the drugs, as a result of some other abnormal brain processes. With LSD, perceptual changes are common, and flashing lights and animals may be seen. On occasion, ordinary objects can become transformed into visions of beauty, and the individual may be overcome by peace and joy. There can be a feeling of religious awareness, which can be expressed in terms of oneness with all of creation. Limbs may feel very long. But there can also be terrifying experiences of unreality. One's body may feel as if it is rotting, and one may see the skull behind another person's face. Experience of death and rebirth, union with the universe or God, and encounters with devils, are reminiscent of similar descriptions in the sacred literature. Experiments with twenty Christian theological students, who were given either a psychedelic drug or a placebo, showed that those who received the drug had profound and positive mystical experiences, which they still valued six months later. A simple drug like LSD could only have such effects if the circuits for these experiences were already there in the brain.

Aldous Huxley's experience with drugs led him to believe that the brain was a screening mechanism keeping out these strange perceptions, and that they could only come in when drugs were present. When he then looked at a vase of flowers he saw '. . . what Adam had seen on the morning of creation – the miracle, moment by moment, of naked existence . . .' Timothy Leary's first experience with taking LSD had a profound influence on his beliefs: 'It was above all and without question the deepest religious experience of my life . . .' And Allen Ginsberg heard William Blake's voice when reading his poems, felt that his body was afloat, and became convinced that he had been born to experience this universal spirit. It is striking how much these experiences have in common – they are not about sex, or food, or work, but are similar to the other delusions and hallucinations just described, particularly those held by mentally ill patients. It is, however, necessary to take into account the personality of those who have such experiences, and whether they are particularly susceptible. The experiences clearly relate to many religious and paranormal experiences, as will be discussed. The similarity of the experiences makes clear that they are an intrinsic property of the human brain. It is most unlikely that any other animal would be significantly affected by these hallucinogenic drugs, as the basic circuits for delusions are not present in their brains.

Hypnosis can give rise to false beliefs, and understanding the processes involved could illuminate key aspects of beliefs that fit so badly with the real world. One reason hypnosis is so important is that it is most often a state induced by the influence and authority of someone else. All the phenomena of hypnosis, the altered experiences, involuntary actions and amnesias, are produced by suggestion. Hypnosis can be thought of as a social interaction in which the subject responds to suggestions that alter perceptions and memories. Yet the subject retains some normal awareness. Hypnosis is not that well understood, but suggestibility is a key feature. Suggestibility could also have a very important influence on our beliefs in our day-to-day lives.

Suggestions made under hypnosis can unquestionably affect physiological aspects of our bodies, as in the classical experiments using the tuberculosis test. In this test, a small amount of the test substance is placed on the subject's arm. If the subject is resistant to tuberculosis, then a small red swelling develops because of the body's immune response. A person with this positive response was hypnotised, and told that there would be no response when his right arm was treated, but that his left would respond. Following local inoculation, there was a red swelling on the left arm, and not the right. But examination of the right arm region showed that the cells responsible for the response had indeed accumulated, but the hypnotic suggestion had prevented changes in the local blood supply and so there was no swelling. There is also very good evidence that hypnosis can result in a reduction of pain, which may be related to the placebo response discussed later.

Hypnosis can also give rise to a delusion, which can again be defined as a belief that goes against all the evidence and which others do not share. Under hypnosis, the subject experiences the conviction that the world is as suggested by the hypnotist, even though it does not conform with reality. It is certainly possible to use suggestion with hypnotised subjects to change their beliefs about themselves and about the world, at least on a temporary basis, and in some instances more permanently. When confronted with evidence that their belief is false, subjects provide what is for them an explanation rather like confabulation.

To investigate this process, subjects were hypnotised to believe they were of the opposite sex. They were then asked what they would say if a doctor entered the room and challenged that belief. They were also asked to look at a video of themselves, and then asked how they could reconcile that image with their believed sexual identity. Highly hypnotisable subjects experienced a change in sex, and one commented afterwards: 'It was so real it was disgusting.' When confronted with an imagined doctor, they argued that the doctor was simply wrong, possibly a quack. They also denied that they were the person on the video, claiming it was a person who had nothing to do with them.

There are also some studies that show that providing subjects with misinformation in the form of a vivid 'reliving' of a fictional past event under hypnosis, in this case hearing gunshots in the night, can create a clear belief that these events have really happened. The belief in this case was resistant to explanations of the experimental, and fictitious, nature of the hypnotic experiences. The creation of 'alien abduction' experiences using hypnotic procedures can be elicited quite readily in even moderately susceptible subjects, with no prior history of alien abduction claims. The experiences produced, however, can be quite powerful and compelling; and if they were presented in a less sceptical context, they could be construed by some subjects as evidence of their own forgotten, or repressed, actual experiences of being abducted by alien beings, with consequent changes in their beliefs on the subject. Some of those who were just observing these demonstrations had their beliefs in the reality of alien abduction confirmed and strengthened by what they saw. They were convinced that the hypnotist had happened by chance on subjects who actually had been abducted, but had forgotten the experience until they were hypnotised. There are doubts whether attempts to change fundamentally held beliefs – such as political beliefs – would be successful, as hypnotic subjects seem to retain a capacity to resist unacceptable thoughts and ideas if they are presented in a confrontational way.

It remains unclear how the hypnotic state is induced, but

high suggestibility is central. What, then, does this tell us about normal belief induction? Do religious rituals contain a hypnotic element induced by the leader?

A hypnotic subject can be persuaded to be so imaginative as to be able to perceive and interact with objects that are not there – to have hallucinations. These hallucinations can include seeing a fly, hearing voices, feeling heat in a cold room, and noticing smells. Negative hallucinations can also be suggested; these include the inability to see one of three balls, not seeing the minute hand on a clock, and selectively being unable to hear the ticking of a clock. There is some correspondence between these hallucinations and those induced by a drug like LSD.

The ability to accept two completely contradictory sets of information at the same time is known as trance logic. The most dramatic examples occur under hypnosis. For example, a subject claims to 'see' two real images of his wife on two nearby chairs, neither of which has his wife on it; and how could there be two of them? Dissociation thus enables the subject to ignore some aspects of reality, so angels can be believed in without any real evidence. Suggestibility is what underlies the process, and is deeply involved in many other beliefs, from religion to psychopathology. One may wonder about the role of authority in creating moral beliefs, an issue to be discussed later.

Hypnotised subjects can behave as if they have seen non-existent events and hear voices that were suggested to them, although no sound was made. They also claim to feel no pain when given a painful stimulus, or lose control over the normal movement of their limbs. It is thus natural to believe that the hypnotic state differs fundamentally from one's normal mental state, and that it involves a special set of mental processes, rather like entering into some sort of trance-like state. But the evidence is far from clear.

There are two main approaches to understanding hypnosis, one being that hypnosis is a special mental state with three main features: it is an altered, sleep-like state of consciousness; the subject responds involuntarily; and the subject can have unique experiences. In this theory, consciousness is seen as split along a vertical axis with multiple systems of control, all of which are not active at the same time. By contrast, other theories see hypnotic behaviour as purposeful action in which the subjects interpret their situation in a particular way. The subject deduces what is expected, and tries to bring about the suggested effects, and the subject's own beliefs about the process can be very important. There is thus no altered state of consciousness. It comes as rather a shock to find that there are theories, quite well established, that totally reject views that in hypnosis the subject enters into a special mental state. Sociopsychological theories of hypnosis view it as similar to a common form of social interaction, basically social compliance. The hypnotic subjects guide their behaviour in terms of understanding the demands of the hypnotist, and their responses are goal-directed. The subjects show willingness to adopt a hypnotic role. Much of the evidence for this comes from having groups of subjects that include 'simulators', that is, people who pretend to be hypnotised. While there are some differences in the experiences of the genuine subjects and the simulators, these can again be accounted for in terms of the genuine subjects' expectations.

Nevertheless it is not that easy to reconcile this view, which explains hypnosis by compliance, with the observations in relation to the tuberculosis test described above. And consider the following very impressive experiment on the modulation of colour perception. Eight highly hypnotisable subjects were asked to look at varying patterns, both coloured and on a grey scale, while their brains were being scanned. In each case, it was those parts of the brain that responded to colour that were scanned. When they were hypnotised, the colour areas of the left brain hemisphere were active when they were asked to perceive colour, even when they were not shown any. Mental imagery cannot activate colour on the left side, and so this is very good evidence for a special hypnotic state, and not just social collusion. Suggestibility can thus underlie many different beliefs, and we may be less in control of what we believe than we would like to believe.

All these examples of distorted and pathological beliefs show how the belief engine can malfunction. These are extreme cases, and it is not hard to imagine that similar distortions could occur on a smaller scale in our everyday thinking, though we try to keep such distortions out. These numerous examples should make clear the complexity of the brain in relation to acquiring, and retaining, false beliefs. There are circuits subject to suggestibility, and others that persuade one that one's beliefs are valid. In all cases, there is an attempt to tell a consistent story, to avoid cognitive dissonance, and so come very close to confabulation. There are clearly circuits in the brain that can, for example, give rise to paranormal experiences when stimulated with quite simple chemicals. Moreover, many of the mental illnesses described have a strong genetic basis. Genes affect how we believe.

We now turn to an attempt to understand the origin and nature of religious and other beliefs, and to the extent, if any, to which they may have given rise to the brain circuits relating to the false beliefs just described.

CHAPTER EIGHT

Religion

... believe in your heart that God has raised him from the dead,
and you will be saved.
The Bible, Romans 10.9

A very common association when the word 'belief' is mentioned is religion. For many, the two are almost indistinguishable. Religious beliefs are universal, complex and variable, and present a difficult problem when we are considering how they originated and are acquired and modified. My suggestion is that they all had their origin in the evolution of causal beliefs, which in turn had its origins in tool use. Given causal beliefs, it was natural for people to ask 'Why' questions about life and death. Since the mental ability that leads to causal beliefs evolved from our primate ancestors and is largely determined by our genes constructing the appropriate neural circuits, one must ask if religion is also partly in our genes. Another question is whether the supernatural beliefs in religions bear any relation to the mechanisms that generate delusions and hallucinations of the type discussed in the previous chapter.

My approach here is to try and understand why there are religious beliefs, and what determines their special character, namely their involvement with supernatural forces; then to look at religion in evolutionary terms, particularly at the possible advantages it gives with respect to health and social life; and finally, of course, to consider whether religious beliefs could be partly genetically determined. I do realise that my analysis is speculative, and my evidence is often weak.

One needs to consider the lifestyle of the earliest humans. They were hunter-gatherers, and group activity and tool use were very important. A key proposal I wish to put forward is that once causal belief evolved in relation to tools, and once language evolved, it was inevitable that people would want to understand the causes of all the events that affected their lives, from illness, through changes in climate, to death itself. Once there was a concept of cause and effect, ignorance was no longer bliss, and this could have led to religious beliefs. People wanted to know what caused the important events in their lives, what would happen in the future, and what action they should take. Uncertainty about major issues that affected their lives was as intolerable then as it is now. There were feelings of fear, of illness and other dangers like starvation, that had to be overcome.

Jacques Monod, the molecular geneticist, and Bertrand Russell have both pointed out how soulless it is to recognise that man is the product of causes that had no pre-vision of the end, and that there is no life after death. William James claimed that 'how to gain, how to keep, how to recover happiness is in fact for most men at all times the secret motive of all they do'. How do our beliefs help or hinder in this plausible scenario? Religion can help, because it promotes optimism and hope. It provides believers with a sense of purpose and meaning in life, but can also give rise to negative effects such as guilt.

First and foremost, our ancestors wanted to know the causes of 'evil' and incomprehensible events. The one causative agent that our ancestors were sure about was their own and other peoples' actions, particularly those learned from tool making and altering the environment. They are the most clear-cut examples of causes that anyone knows, and are a clear example of how representativeness can mould our beliefs. Such beliefs may have been supported by delusions and dreams involving unknown humans. So it should come as no surprise that so many of the answers to our ancestors' causal questionings should have human qualities together with powerful forces. For many religions, there is a belief in a god who is like a person without a normal body: free, eternal, all-knowing and capable of doing anything. This god is the proper object of worship, and can return the world from evil to normalcy. A number of faiths have a strong similarity with regard to the idea of creation, as well as physical cause. The key notion is that the world was not only created by some divine being, but is being continuously maintained and changed. For example, life itself was created by gods, as was the rest of the universe.

There are so many religions and the differences in detail with regard to their beliefs are so great that they can contradict each other. In this sense most must have beliefs that are wrong, though there will be many who will argue for the validity of one religion compared to all the others. Pascal Boyer, an anthropologist, recounts a nice story about telling the other diners at a Cambridge college dinner how the Fang people believe in witches who have an internal organ that can fly at night and destroy crops. There were even claims of some of the Fang having seen them in flight. To this, a prominent Catholic theologian said that he wondered how it could be possible to explain such nonsensical beliefs. But as Boyer knew, the Fang were themselves totally puzzled by Christianity, particularly by the Trinity, three people in one, and by the belief that all misfortunes were due to two ancestors eating some exotic fruit.

Religion may thus be thought of as one major way of finding meaning and value in the difficulties of daily life. Religion is practical, as it almost always involves interaction with, and help from, some supernatural being. All religions have some beliefs about death and the fate of the dead, which may be reassuring to the living. Religion can also be viewed as a tradition transmitted by teaching, whose fundamental feature is its ability to help people deal with the problems of life, while at the same time by facilitating social cooperation and providing a sense of identity. Jews, for example, believe that they were the chosen people. Religious beliefs provide answers to difficult questions, and can give order and meaning to situations even where explanations are absent. Religion provides an explanation for the causes of evil events, and this helps to maintain religious observance. For the Muslim believer, the Koran is the supreme book of divine revelation that derives from the Almighty and reveals God's eternal word.

New religious movements often arise when there is some sort of serious social change that does not easily fit with traditional beliefs. They often offer answers to questions addressed by traditional religions, like the purpose of life and the nature of death. Yet the diversity can be enormous, combining ideas from Christianity, Islam, science fiction, psychoanalysis and Satanism. There is often a leader who wields a charismatic authority.

There are very few societies without religious beliefs. Almost everywhere, the ways in which people meet the problem of death are in the realm of the sacred. One of the earliest ideas about the origins of religion comes from the British anthropologist Edward Tylor, who in 1871 proposed that all religion is a belief in spiritual beings arising from the experience of death, and from dreams together with hallucinations. There have been numerous discussions on this theory and though we will never know for sure, it seems plausible. One can note, for example, the importance given to ancestor worship – there is evidence of burial ceremonials going back at least 60,000 years.

It is generally assumed, in almost all religions, that there is an afterlife: Heaven, Hell, Valhalla, Nirvana. For our ancestors, this made sense, for where did the person go when the body was lifeless? It could not just disappear. Almost everything in their environment had some permanence unless they could see the cause for its destruction and disappearance. They thus imagined the existence of the soul, which also provides a way of dealing with the fear of death, for the soul can live on. It thus made sense to honour, even obey or blame, one's dead ancestors. All this could have reduced the fear of death, and so have been an evolutionary advantage.

Pascal Boyer has considered the question of what makes religious beliefs seem so natural to many people. His answer rejects the view that it is just people's wish to deal with misfortune or understand the universe, though he recognises the importance of gods reacting to one's behaviour. For him there is no simple answer; a variety of cognitive processes are involved, which are used to account for evil events. He argues that an important aspect of religious notions is that they are products of the supernatural imagination, which in turn involves counterintuitive notions. These violate certain expectations and confirm others. Gods are persons with extraordinary powers; their statues are inanimate but can hear one's prayer, and ghosts have some of the properties of a person but can go through walls. Some people in the Sudan believe that ebony trees have the capacity to eavesdrop on conversation near them, and may reveal what they have heard. A shaman burns tobacco leaves in front of a row of statuettes and asks them to go and cure a friend whose mind is being held hostage by invisible spirits; a witch can hit a person with invisible darts and poison their blood; an animal is sacrificed in a particular way to appease dead people. These are typical of a wide variety of supernatural religious causal beliefs.

Buddhism, in terms of belief, appears to be a religion. But, and this is crucial, there is no God. Buddha was an actual historical person and did not claim to be divine, but because of his inquiry into the nature of the world and self, he achieved enlightenment. Enlightenment is not meant to be mystical, but the way of clear-seeing rationality, and this is possible for any individual. By activating your Buddha-nature, it is suggested, you too will achieve enlightenment. In the *Upanishads* there is the claim that there can be a new dimension of self, which is the same holy power that sustains the whole world. From this, Buddhism emerged, and the Buddha taught that it was possible to gain release from the sufferings of life by essentially good and compassionate behaviour, combined with the sense of transience in meditation – clearly a causal belief. Buddhists view existence as a series of stages endlessly repeated, and there is reincarnation, but not necessarily a human-like afterlife. In Tibetan Buddhism, the Dalai Lama is believed to be an incarnation of the Buddha of Compassion; when the Dalai Lama dies, his soul is believed to pass into the body of an infant.

Religion is concerned with the supernatural, and thus involves forces and causes beyond our normal experience of nature, and this is something we need to understand. Humans will perhaps seek rewards through the supernatural if they are not obtainable by other means. Religion consists of very general explanations of existence, including the terms of exchange with a god or gods, for example, for good health. And since causal beliefs that promote survival are partly programmed by our genes, could that not also be true of some aspects of religious beliefs that promote survival, particularly those that relate to mystical forces, and even, perhaps to the gods themselves? In addition, religious beliefs provided gods or ancestors who could be prayed to, and who might help to solve problems. Again, those with such beliefs may have been better adapted for survival if they were less anxious and healthier. The evidence for this is not strong, and religion and health will be looked at below.

Religious experience often involves an approach to God or some other mystical presence that induces awe, or the experience of unity with God and a feeling of peace and being at one with the world. It is not easy here to distinguish between belief and emotion. There is no serious regret by those who have such experiences that they cannot fully describe their experience to others. The experience of intense emotion in a non-religious setting may be quite similar to a religious one. This may reflect the altered states of consciousness and hallucinations that are part of many cultural practices worldwide, and so may also be related to suggestibility and hypnosis. There is also the suggestion that religious experience is the result of creatively solving a personal problem relating to a crisis of identity.

Religion is almost always regarded by its believers as a way of obtaining help from supernatural powers, possibly from a god. Miracles can win further adherents, and the Bible has many examples, not least the dividing of the Red Sea to allow Moses and the Jews to cross. However, as David Hume argued, no miracle should be believed in unless the evidence was such that it would be miraculous not to believe in it.

Belief in a religion is not all or nothing. While there are some believers, for example, who may feel God is always present, for others, religion matters only on special occasions like weddings, births and funerals. There is thus a continuum between disbelief and belief in religion. That religions can be a source of comfort was claimed by Karl Marx, who spoke of religion as 'the sign of the oppressed nature', a view expressed by many others, including Nietzsche, who said 'Christianity is a sickness, arising from the envy of the underprivileged who felt it justified their position.' About half of the Christians asked in a survey believe that Jesus Christ will return to the earth one day. Such messianic beliefs are not restricted to Christians, and the idea of a messiah is much more widespread. Millions of born-again Christians in America believe that an event called the Rapture is coming, and Christ will return.

Believing that God is in control does not mean that people believe that they themselves have no control. The evidence is that they believe they do have control over their lives, and that God and prayer provide an important set of additional tools. They also believe less in chance governing their lives. There is good evidence for a positive correlation between being religious and being happy. This may in part be the result of assigning God as the cause in matters relating to health and death. Prayer is very important because the individual believes that he or she really can influence what will happen; some people believe they are empowered to communicate directly with the source of all control and change. Most religions teach that suffering is to be expected and could even be a valuable experience. Meaning is essential: if the meaning of suffering is clear, it is easier to bear. Death needs an explanation, and religion can provide it. We go to heaven or hell, our shadows persist, we become ancestors. Most religious explanations are comforting, but not all. Nevertheless, religious belief is a way of dealing with adversity in the modern world; soldiers find prayers helpful before battle.

All societies have beliefs about the origin of the world and what happens after death. Religion has declined a little in the West, but is still a powerful force in both the West and the East. An ICM poll in 2004 found that 85 per cent of Americans believe that God created the universe, while in Britain only 46 per cent believed in God. In Nigeria, 98 per cent claimed always to have believed in God, while nine out of ten Indonesians said they would die for their God or religious beliefs. A survey by the market research bureau of Ireland found 87 per cent of the population believe in God. Rather than rocking their faith, 19 per cent said tragedies such as the Asian tsunami, which killed 300,000 people, bolstered their belief. In a 1996 Gallup poll in the USA, 90 per cent believed in heaven, 79 per cent in miracles and 72 per cent in angels. According to a 1991 Gallup poll, also in the USA, over one half believed in the devil. During the last thirty years, membership of the Mormon church has more than doubled to around 5 million, and that of the Southern Baptists has increased from around 8 to 16 million.

In 2001 just over three quarters of the UK population reported having a religion. More than seven out of ten people said that their religion was Christian (72 per cent). After Christianity, Islam was the most common faith, with nearly 3 per cent describing their religion as Muslim (1.6 million). Religion in the UK has declined in the 20th century as measured by church membership. Only the Catholic Church showed an increase. In 1998 about half of the people of Britain believed in God, but even occasional church attendance was down to about 25 per cent. It is not clear why, if religion is in our brains, so many do very well without it. The evidence does not point to science, but rather to the emergence of complex technologies and an industrial society where prayer and religion play a less clear role.

Two themes dominated social scientific theories of religion for some three centuries: the first that gods are illusions generated by social processes, 'society' as the nineteenth-century sociologist Emile Durkheim put it, 'personified and represented to the imagination'; the second that the gods are illusions generated by psychological processes that are the product of a primitive mentality in so-called 'primitive' societies. There was in the nineteenth century a quite widely held view that 'primitive' people had a primitive mind, with lower intelligence compared to their Western observers. The Victorian sociologist Herbert Spencer even went so far as to say that they had no concept of cause, and were without curiosity. This primitive mind hypothesis has been used to discredit not only the mode of thought in 'primitive' societies, but religion in general. But the scientific study of religion by social scientists is relatively recent. To his credit, at the beginning of the twentieth century the French philosopher Levy-Bruhl did recognise the similarities in 'primitive' thought in all cultures. But it was the Polish anthropologist Malinowski, from his personal studies of the Trobriand Islanders at around the same time, who made it clear that they were well capable of rational thought, and only resorted to supernatural beliefs as a last resort. They worked their fields perfectly sensibly, but controlling the weather was something for which they had to turn elsewhere.

How does belief in religion come about? There is, in general a set of copy-me programmes involving the strong influence of a religious community. For children, religion is learned from others, particularly their parents. A personal religious experience can be very important. William James, in his wonderful book *The Varieties of Religious Experience*, tried to understand religious belief and feeling in scientific terms; to lay bare its causes, following both Spinoza, who said that one could analyse the appetites of men as if they were a question of lines, planes and solids. James's book is as important and relevant today as when it was written over a 100 years ago.

James sees religion as the feelings, acts, and experiences of individuals in dealing with themselves, and how they stand in relation to whatever they may consider to be divine. He is thus concerned with individual experience rather than the institutional features of religion. He regards melancholy, for example, as essential to every complete religious transformation, in addition to the happiness that religion can confer. However, he is not sure that these emotions are different from the same feelings when not associated with religion, though he sees solemnity as an essential feature of the divine. Religious experience, James argues, is as real to those who experience it as an experience of the sensible world. There is thus nothing in the approach of any rationalist that could lead one to deny the 'reality' of such experiences, for 'the extent of disbelieving in certain types of deity makes theologians of us, for these disbeliefs are a theology'.

Strange experiences, not necessarily religious, but with a related spiritual flavour, are a common occurrence, as we have seen. Reports claim that over a third of adults in England and the USA have experienced them. These experiences might be hallucinations or delusions, and they have a strong emotional content. James doubts whether there would have been the myths, dogmas or superstitions that characterise religions without those personal mystical experiences.

What triggers a religious experience? A survey of subjects who felt they had been close to a powerful spiritual force showed that there were a variety of triggers, the most common being music, the beauty of nature, children, poetry and sex. But for soldiers, and others, stress and danger and extreme fear could provide the triggers. These feelings can be very difficult to describe; they can involve a sense of standing outside oneself, or extreme feelings of love and beauty. Some triggers could operate by stopping the flow of thought so the mind becomes very narrowly focused, or involve special changes in brain function. One possibility is that religious experiences are related to the delusions that can occur when false beliefs are activated. Trying to reach different states of consciousness is central to religious traditions, as in, for example, meditation. We have noted how mental illness and drugs can generate religious experiences. Special techniques have been developed to assist entrance into such states, particularly deprivation of food or sleep, or control of breathing. There are reports of individuals who have near-death experiences that can involve a dark tunnel with a light at the end; a review of one's life; feelings of peace and serenity; even coming back to life again. Near-death experiences were already described by Plato, with the tale of Er, who returned from heaven with a description of the visions he had had, including a bright column of light. In a 1982 poll, 15 per cent of adult Americans reported having had a near-death experience, and an earlier survey in China found the number as high as 40 per cent. Usually, pain and fear were replaced by calm and a sense of peace.

The apparent reality of dreams shows that our brains can unconsciously generate experiences that have not actually occurred. Dreams may have played a role in creating certain religious beliefs. They can give an uncanny sense of a real event in a way that is still poorly understood. The Assyrian epic of Gilgamesh, written around 3000 BC, records dreams that are interpreted as messages from the gods. Babylonian writings also contain recordings of dreams. So, too, does the Upanishad from India. And in the Iliad and Odyssey they play an important role, as Zeus sends a dream figure to Agamemnon to urge him to attack Troy, and Penelope dreams about Odysseus' return. Aristotle, long ago, recognised that many people supposed dreams to have a special significance.

Religious concepts can be used by people when there is a need for them. They are used to account for a particular occurrence, like someone's death or an accident or a drought. For example, the Kwaio in the Solomon Islands believe that good crops show that the ancestors are happy with the way they are behaving. While ancestors play a key role in determining their fortune, they are very vague as to where they might live or how they exert their influence. This is common: just how religious agents perform their good and bad works is rarely a matter for reflection or interest.

To what extent are humans rational and religious beliefs

irrational? The psychology of religion is permeated with arguments attributing it to psychopathology, groundless fears and
faulty reasoning, and showing it is irrational. William James, in
his essay 'The Will to Believe', provides a justification of faith, a
defence of our right to adopt a believing attitude in religious
matters, in spite of the fact that our logical intellect may not
have been persuaded:

> Since belief is measured by action, he who forbids us to believe
> religion to be true, necessarily also forbids us to act as we should if
> we did believe it to be true. The whole defence of religious faith
> hinges upon action. If the action required were inspired by the
> religious hypothesis this is in no way different from that dictated by
> the naturalistic hypothesis, then religious faith is pure superfluity,
> better pruned away, and controversy about its legitimacy is a piece
> of idle trifling, unworthy of serious minds. I myself believe, of
> course, that the religious hypothesis gives to the world an expres-
> sion which specifically determines our reactions and makes them in
> a large part unlike what they might be on a purely naturalistic
> scheme of belief.

Can our will, James asks, really help or hinder our intellect in
its perceptions of truth? He analyses beliefs in terms of gains and
losses, cost benefit, and considers, for example, Pascal's wager
that you must either believe or not believe that God exists. A
game is going on between you and the nature of things, which at
the day of judgement will bring out heads or tails. If you win you
will gain beatitude, and if you lose you lose nothing at all, and
therefore it is worth taking the chance of believing.

John Habgood, a former Archbishop of York, writes in his
book *Varieties of Unbelief* that belief and scepticism are essential
components of serious thinking. For him, unbelief is based on
ideas that its believers claim are more secure than they actually
are, for God is essentially unknowable, so trying to put into
objective terms what can only be known from within opens it

up to unbelief. Again, for the physicist Russell Stannard, the reality of God stands or falls on the perceived 'otherness' of religious experience. Moreover, for him there are similarities between science and religion, in that both try to discern patterns of intelligibility lying behind what at first seems to be a bewildering array of contingent events.

Michael Shermer conducted a survey of the Sceptics Society in the USA, which has many members who are scientists, and found, to his surprise, that while they were indeed sceptical of the paranormal, over one third thought it likely God existed. He then tried to find out the basis for this belief, not only among them, but more generally. The answers he received were of two main kinds: first, that there appears to be a pattern of God's presence in the world, and secondly, that belief brings comfort and alleviates fear of death. The latter reason was the main explanation for those questioned not about their own beliefs, but about why they thought other people believed in God. For those who did not believe in God, some 40 per cent said that it was because there was no proof of God's existence. Other reasons were that such a belief was absurd and unnecessary. It is striking how little a role science plays in such explanations, and though there was some correlation between disbelief and interest in science, this may merely reflect higher educational qualifications. The more educated, the less religious belief. Conservatives were more religious than liberals. And people who scored highly in openness were less religious. Men tried to justify their belief with reasons, women with emotion.

Not to have a religious belief is relatively modern. Unbelief is most likely a product of Greek thought. In his Laws, Plato argues for religious beliefs: the existence of God, and the immortality of the soul. Not to believe in these he regards as a serious crime worthy of prison, even death. For him, it is not a matter of acceptance but faith, trust and obedience. Tertullian, however, in the third century AD, claimed 'The son of God is dead: this is to be believed because it is absurd. Having been buried he rose again: this is certain, since it is impossible.' The Bible's key ideas were not open to question, and until the eighteenth century non-belief was limited to a relatively small intellectual group. But even they recognised the powerful social influence and the social utility of religion. They appreciated the increase of objective belief based on reliable evidence has not led to a collapse of religion. As the American poet Wallace Stevens puts it, 'We believe without belief, beyond belief.' Or W. B. Yeats: 'Man can embody belief but he cannot know it.' Faith, a la the German theologian Tillich, is 'belief in the unbelievable'. Religious belief is partly the relation of self to the rest of the world in some mystical manner, rather than to just objective reality. Three centuries ago, the radical English philosopher Thomas Hobbes, in his *Leviathan* dismissed all religion as 'credulity', 'ignorance' and 'lies', and said that God existed only in the minds of the believers. He wrote:

> For if a man pretend to me that God hath spoken to him supernaturally and immediately I make doubt of it, I cannot easily perceive what argument he can produce to oblige me to believe it. It is true that, if he be my Sovereign he may oblige me to obedience, so as not by act or word to declare I believe him not; but not to think otherwise than my reason persuades me . . . For to say that God hath spoken to him . . . in a dream, is no more than to say he dreamed God spoke to him . . .'

Voltaire was surprised by the presence of widespread atheism in England in 1730 and David Hume wrote his famous essay on miracles in 1748. 'There is not to be found in all history any miracle attested by a sufficient number of men, of such unquestioned good sense, education and learning, as to secure us against all delusion in themselves . . . It is strange that such prodigious events never happen in our days.' Other attacks on religious beliefs came earlier from the French historian of mathematics and science de Fontenelle in 1687, and as early as 1711 the Earl of Shaftesbury was putting forward a scientific study of religion, which he discussed in terms of anxiety and illusion. Such an approach is at the start of a now long tradition of attributing religion to groundless fears and irrationality. Yet at those times in Europe, insanity was believed to be either due to the devil, or a punishment by God.

Sigmund Freud described religious beliefs as 'illusion' and 'childishness to be overcome'. He developed a psychoanalytic theory of religion based on the Oedipus complex: the desire of the son to sexually possess the mother and exclude the father. His explanation, that it had its origin in the killing of a dominant father who hounded females, and in the subsequent guilt of the murdering sons, cannot be taken seriously. He later developed his theory as follows:

> Religion would thus be the universal obsessional neurosis of humanity; like the obsessional neurosis of children, it arose out of the Oedipus complex, out of the relation to the father . . . religion brings with it obsessional restrictions, exactly as an individual obsessional neurosis does; on the other hand it comprises a system of wishful illusions together with a disavowal of reality, such as we find nowhere else but . . . in a state of blissful hallucinatory confusion.

– a description that in some ways could be applied to psychoanalysis itself.

A rather different approach to religious beliefs from that considered so far emphasises the social aspects of belonging to a religious community, and the extent to which this brings advantages to the members. Sloan Wilson looks at human society as an organism in its own right, and thus at the evolution of society in Darwinian terms. This approach is flawed, since there is no way in which a society, religious or not, conforms with the evolution of organisms, as there is nothing equivalent to replication of the genes or their programming of the behaviour of the organism. Moreover, the same ideas could be applied to communism, and Wilson never accounts for the supernatural nature of religion. Religion has various dimensions: doctrine, myth, ethical teaching, ritual, experience, and social institutionalisation; but so have Maoism and Stalinism and Communism. With Maoism, for example, comes the doctrine of Marxism and myths based on the Long March. But is Maoism really like a religion? It seems not, as the supernatural aspect and hidden forces are absent. Nevertheless, Wilson's ideas are interesting in that he considers that religion is more than just supernatural beliefs about gods, and that it provides advantages to those who share certain religious beliefs, and act in accordance with them.

Such an approach, in which one considers the group or an individual as having characteristics that are adaptive – the coat of the polar bear, for example – is called functionalism. Darwin himself came close to a functionalist approach for groups, giving as examples the high standards of morality in a tribe when they help each other and show courage, obedience, and the willingness to sacrifice themselves for the common good. Evans-Pritchard's ideas are similarly functionalist in essence, and can explain the similarities between the Nuer religion in Africa and Judaism as a case of convergent evolution. Both were determined in cultures whose living depended on herding cattle, and both are very humble before God. The Nuer believe that to be right with God it is necessary to be right with one's fellow citizens. One might have thought that a central feature of all religions was a belief in an afterlife, as in Christianity, but this is not the case. The Nuer, for example, care little about what happens after death, although they fear it. And Judaism too gives little attention to an afterlife. So, are religious groups adaptive units in this sense? It does seem possible.

A key problem with the cooperation of individuals in a group is that there may be defectors, those who exploit the goodness of others for their own ends. Hunter-gatherer groups are egalitarian, as their possible selfish impulses are controlled by other members of the group. It is also clear that hunters will do better if they hunt together rather than as individuals. The Chewung in Malaya are egalitarians and are hunter-gatherers as well as farmers. Distribution of food is governed and determined by belief in punen, which is a misfortune caused by not having satisfied an urgent need. To avoid punen, all food is shared out equally, and there is a ceremony to ensure that everyone knows this and so avoids punen. Being in punen can, it is believed, lead to an attack by one or another animal. This system of beliefs emphasises group sharing. The adaptive advantage seems relatively clear, as its effects are clearly related to survival.

Sloan Wilson's attempts to understand a religious community in relation to its environment from an evolutionary perspective focuses on John Calvin's views of Christianity in Geneva in the sixteenth century. How adaptive, he asks, was Calvinism for the inhabitants of Geneva? Calvin placed equal emphasis on people's relationship with God and on their relationship with other people: duties of charity owed to one's neighbours. Calvinism included, of course, the Ten Commandments, which may help societies to have adaptive belief systems that lead to behaviour similar to cooperation with one's neighbours. Again, the system must cope with the problem of some individuals exploiting others' good behaviour. A belief system must be easy to understand – not, for example, based on hard science. It may be more motivating if one regards one's enemy as being inhuman, or bewitched, or intrinsically evil.

Calvin's catechism is claimed to be devoted to establishing such a system of beliefs, and these may be interpreted as

designed to motivate good social behaviour. God–person relationships can reinforce such social cooperation. But it remains
far from clear whether Calvin's society was adaptive in the evolutionary sense of ensuring the reproduction and survival of its
citizens. Its administration may have been very fair, and individuals may have liked many aspects of it, but from an evolutionary viewpoint this is irrelevant, since only reproductive
advantage matters. A better example, perhaps, is provided by
the Balinese goddess who helps with water irrigation of the rice
fields and ensures cooperation in the distribution of the water.
This coordinates the activities of thousands of farmers for their
mutual benefit. In the light of this thinking, Wilson suggests it
is plausible to argue that we are genetically programmed to
have a psychology sympathetic to the adaptive rules of religion.

Another approach has been to view religion in terms of economic theory, that is to see it as an exchange between the
believer and a supernatural being, usually a god or spirit in
human-like form, for something that is very scarce or desper-
ately needed, like rain or a cure for illness. This theory of Stark
and Finke about the social aspects of religion starts with the
proposition that humans attempt to make rational choices, and
is known as the rational choice theory of religion. Religion does
not deal with things that humans can do or acquire relatively
easily, like getting to work or making things. Religion offers
rewards for things that are very difficult to obtain by other
means. This can involve, for the individual, weighing the costs
against the benefits. Thus, their theory seeks explanations as to
how humans try to get rewards and avoid costs.

An analysis, related to this theory, looks at the rise of the early
Christian church. A typical Roman city was a very unpleasant
place to live in – dirty, crowded and dangerous. Antioch, for
example, was plundered on five occasions. Fires, too, were frequent. By contrast, it is suggested, Christian society would have

appeared attractive, for it was easy to become a Christian and those who did became part of a much more close-knit and supportive community. As with Judaism, the Ten Commandments promised cooperation and fairness. They also supported marriage and large families, so Christian women raised more babies than their pagan compatriots. Evolutionary advantage at last. In addition, during infectious epidemics like measles and smallpox, Christians have been described as providing significantly more help to their fellow believers. In a way, Christianity provided something like an early welfare state.

What is the relationship between religion and health? If it is positive, then religion could be an adaptive in evolution. There are intimate connections between health and religion, which will be considered further in a later chapter. Here, only the advantages are considered. There have been extensive studies in this area. Religion was distinguished from spirituality on the basis that it involved rituals, and the studies dealt almost entirely with Western religions. Although the studies should be regarded as tentative, the evidence is that there is an inverse relationship between pain intensity, and religious beliefs. This is consistent with the findings that those within a religious community enjoy better mental health, possibly due to social support. There is also evidence that religious activities reduce psychological stress and promote greater well-being and optimism, and so help to reduce the bodily effects of stress such as that on the heart. Religious beliefs are complex; social religiosity, for example, is associated with a reduced risk of psychiatric illness and drug taking, while other beliefs such as God as judge and general religiosity are only associated with less risk of drug abuse.

Has, then, religious belief a genetic component? The Minnesota twin study did find that there was a genetic influence on whether an individual developed religious beliefs. The heritability was around 50 per cent, which implies a significant genetic component. Almost every culture has a belief in a spiritual world that contains a god who can be prayed to and is in control of powerful forces. The predisposition to religious belief is itself a complex and powerful force, and is probably an essential and ineradicable part of human nature. But it was the psychoanalyst Karl Jung who made the important observation that not only did all cultures possess mythological beliefs, but there were remarkable similarities. He concluded that these beliefs must be inherent in the human mind, and he called it the 'collective unconscious': an early case for genetic determination of such beliefs.

I think that religious beliefs were adaptive for two main reasons: they provided explanations for important events, and offered prayer as a way of dealing with difficulties. Those with such beliefs most likely did better, and so were selected for. As argued throughout this book, our brain has a natural tendency to find consistent and reasonable explanations for important events, and so religious beliefs are most likely partly genetically determined. They are linked to our need to seek causal beliefs, and our minds are largely fashioned by genes specifying how our brains work. There is a tight linkage between genetic evolution and cultural history, and gene-culture evolution has created many human societies with religious beliefs.

Religion may thus be deeply rooted in our biology. There are even claims that link spiritual and religious experience to the activity of specific regions of the brain. A variety of brain-imaging techniques have been used. One model proposes that activation of the autonomic nervous system – the one that is not directly under our control and controls that our heart rate and blood flow – acts on those regions of the brain responsible for mental experience, such as the temporal lobes. These lobes are thought to modulate feelings and emotions. Evidence for a role of the temporal lobes in religious experience comes from epilepsy originating in these lobes, and their association with sudden religious conversions. It is suggested that the visions of St Teresa may have been associated with temporal lobe epilepsy.

It is thus of great interest that the psychologist Michael Persinger has stimulated the brains of subjects with electromagnets that cause tiny seizures in the temporal lobes. Many subjects had supernatural spiritual experiences, even religious ones, which included the sense of something or someone else in the room, distortion of their bodies, and religious feelings. A possible explanation is that a shock can so disturb the system that strange experiences like these result, and this suggests that some aspects of religious experience are programmed into our brains. However, another study failed to confirm his findings.

Certain religious experiences are similar to some false beliefs like schizophrenia and other causes of hallucinations. Having examined the origin of religion and its probable advantages to the believing individuals, I support the idea that religious beliefs have a genetic component. Freud did propose that religious belief reflected an innate desire. This genetic contribution may have programmed our brains to have spiritual and paranormal experiences easily, a view discussed further in the next chapter. When viewed in this way, hallucinations and delusions may reflect a basic programme in the brain that, for a variety of reasons, could be activated at inappropriate times. Thus, religious experiences may have become linked to paranormal beliefs, delusions and hallucinations. It is part, perhaps, of a fundamental need to provide people with beliefs that enable them to understand the world and so determine how to behave, and all this came from the causal beliefs related to tool making and use. The paranormal is considered next.

CHAPTER NINE

Paranormal beliefs

For what a man had rather be true
he more readily believes.
Francis Bacon 1561–1626

A large number of people hold paranormal beliefs even though others claim, strongly, that they are false. These beliefs are all different, but share the property of being in conflict with scientific knowledge, as well as going against everyday experience of causes, and all have a somewhat mystical quality. But more importantly, since they invoke forces and causes outside both ordinary experience and science, they offer believers new powers. Paranormal beliefs include that it is possible to contact the dead; that it is possible to access past lives through hypnosis; that horoscopes can provide useful information about the future; that spiritual healing can cure where conventional medicine fails; that it is possible to transfer thought through telepathy; that angels and ghosts exist, as do monsters and aliens; that it is possible to see what is in someone else's mind; that spirits exist, and can move objects; and even that levitation is possible. It is also not unreasonable to think that some religious beliefs are paranormal – consider Christ's miracles, his rising from the dead, and the supposed effectiveness of prayer. Paranormal beliefs may partly be the result of trying to interpret events for which no simple explanation seems possible, together with a preference for invoking mystical causes and forces.

As has been discussed, our brains contain a belief-generating machine, an engine that can produce beliefs with little relation to what is actually true. Many of our beliefs are not based on evidence that we have examined, but on what we have been told by 'authorities', and on what an individual might have experienced as being quite out of the normal – something mystical. Coincidences are often over-interpreted, and many people have coincidences that they interpret as telepathy such as knowing which friend was on the phone before they picked it up. There are beliefs about forces, causes and phenomena that it is not possible to replicate, but that open up new possibilities. There is also a general lack of understanding of science and of how to judge evidence about causes that go against science. Socrates, in *The Republic*, suggested that it might be possible for a ruler to introduce a myth to support his own position; that, for example, he was made from gold, while the masses were made from iron. Would people believe it? Perhaps, he suggested, not at once, but very likely in later generations.

It is also important to recall the evidence from false beliefs to see how our minds seem to have been programmed for mystical and paranormal beliefs, possibly linked to religion. Just recall how taking psychedelic drugs can lead to very strange beliefs. Schumaker refers to the 'paranormal belief imperative'. He argues that we are pre-eminently autohypnotic creatures and suggests that human beings are 'a believing phenomenon, who must believe in order to live at all'. Others have suggested that a full apprehension of the human condition would lead to insanity, and that paranormal beliefs can help relieve this stress, particularly the belief that we live in some domain larger and more permanent than mere everyday existence. Moreover, for many, science is inaccessible and unsatisfying.

Witchcraft, astrology and magic flourished in the sixteenth and seventeenth centuries in England. One of the central features of all three was a preoccupation with the causes of human misfortune. One such cause was the short life expectancy, which for males was around thirty years; others were that the food supply was unreliable, and disease common. Bubonic plague was rife and almost as dangerous as fire. Astrology was important, and some claimed it was the only way to discover witchcraft. These paranormal activities ran in parallel with religion to help with daily problems. There was a common belief that suffering was due to a person's moral failing; ghosts could torment those who had failed in their moral duty. If cures failed, there was always an excuse offered by the wizard, usually blaming the patient for not following instructions precisely.

Hobbes, in his *Leviathan* (1651) attributed witchcraft to ignorance of how to distinguish between dreams and experiences when awake. But every author in Restoration England believed in the supernatural. Magic, according to Voltaire (1756) 'is the secret of doing what nature cannot do. It is an impossible thing – superstition is, after the plague, the most horrible flail that can infect mankind.' And again, 'The church has always condemned magic but has always believed in it. Magic rose in the East and then became enshrined in Christianity.' As late as 1782 a witch was tried and executed in Switzerland.

In 1857 there was a scandal in the French countryside as an epidemic of possession broke out in the Haute-Savoie. People hurled insults, swore and prophesied, and received messages from outside spirits. Only after fifteen years did the police put an end to the chaos. Priests were sometimes accused of being responsible for the outbreak of illnesses like cholera. At about the same time, a vicar in North Yorkshire was noting how persuaded he was of the 'very real and very deep seated existence of a belief in the actuality and the power of a witch' among his parishioners. What is so important about witchcraft and other related beliefs is that they provide an explanation for important events that affect people's lives.

Supernatural agents can be very different: there can be one supreme God or many spirits and ancestors. And as Evans-Pritchard made so clear some 50 years ago, many of the magical beliefs of the Zande in Sudan were sensible. They knew that termites could cause a mud house to collapse and injure the inhabitants – but what witchcraft could explain was why it should be that particular house, at that time, and with those people inside. It was the particular event that was so important. Again, a young man running through the forest trips and hurts himself. Yes, he knows he did not look properly at the ground, but why did he not do so? Witchcraft. The Zande had no problem with believing in natural causes – they understood how things happen, while witchcraft explained why they happen. This distinction is fundamental.

It is possible to see here some similarity between witchcraft accounting for things that go wrong, and our Western belief in luck, or in bad luck. Witchcraft is not responsible for social behaviour like lying or adultery. Breaking of taboos, like having sex before the child is weaned, and the consequent illness of the child, is not witchcraft. It is when common-sense explanations fail that witchcraft is invoked. Indeed, all bad events have a similar or related cause. Yet there is something of a contradiction in Zande thinking, for they pleaded innocence of witchcraft themselves, yet accused others of it. They could also be sceptical about witch doctors, and often claimed they were frauds. Yet they were so certain about their beliefs that they would not consider doing anything to test them. When an oracle failed, they just found excuses. They do not postulate any mysterious spiritual being, but only the mysterious powers of humans. Their belief in witchcraft is not all that different from conspiracy theories of history, or the supposed power of psychoanalysis to cure mental illness.

For the Zande, witchcraft played its part in every activity, from a wife being sulky and unresponsive, to blight of the groundnut crop, or the cracking of a pot being made by an experienced potter. Similar views have been expressed by the historian Robin Briggs in relation to his study of witchcraft in the seventeenth century in Europe. Very few people are content to accept that blind chance plays a large part in their lives; they seek reasons and logical connections even when these do not really exist. The human mind, far from being infinitely malleable, tends to impose certain inbuilt patterns on experience. The presence of strikingly similar witchcraft beliefs in most known societies raises the relationship between witchcraft and human universals. Briggs asks whether human beings are born with a specific inherited mechanism for detecting witches somewhere in their minds, which excites them when activated. Belief in witchcraft provided intuitively attractive ways for evading logic, as there was no real evidence for the witches' crimes. Witchcraft was about envy, and the power to harm others.

How then did such a system change and magical beliefs decline? In part there was the rise of science in the seventeenth century, which may have weakened the belief in miracles. Technological advance in the eighteenth century probably helped remove some magical beliefs; there was, for example, an increased sense of an orderly routine with the growing use of watches. Also, the practices related to witchcraft were found to be unsatisfactory when they failed to work. The industrial revolution, with its machinery, was also a factor: as the environment came more under the control of technology, conditions improved.

Yet there was some emancipation from magical beliefs before there was any effective technology to affect them. It was perhaps the corrosive impact of social disruption. Perhaps urban living and the idea of self-help were important causes. Urban culture, newspapers, schooling, the army, all 'crumbled the cake of custom'. However, the average men and women of today, while rejecting the evidence for witchcraft, would still lack any reasonable explanation of the phenomena that witchcraft explained.

Levy-Bruhl's statement is very important in this connection:

> From a strictly logical point of view no essential difference has been established between primitive mentality and our own. In everything that has to do with everyday experience, transactions of all sorts, political life, economics, the employment of numbering etc., they behave in a way that implies the same use of their faculties as we make of our own.

But he went on to say that:

> There is a mystical mentality more marked and more easily observable among primitive societies than our own, but present in every human mind. That is a sense of an invisible power and a reality other than our normal reality. It is something fundamental and indestructible in human nature.

Like Jung's claim, a case is made for genetic determination of mystic beliefs.

Edward Tylor was an early thinker on the reasons why belief in the paranormal was near universal across history and different cultures. He argued that people were from the earliest times deeply puzzled by two phenomena: the difference between a living and a dead body, and the nature of the people in dreams. This led to the belief that life could leave a body and go wandering, as it does in dreams. Even when one died, a ghost might still exist, and even appear to others. Communication with the dead thus became feasible, and this possibility of ultimate immortality was, and still can be, very comforting. The visions of dreams may be related to the 'spirit' that leaves a dead person.

Sleep is itself a sort of paranormal experience, and dreaming of a dead person would make it more so. And illness, in early times, must also have been almost a paranormal experience, as would be childbirth, the causes being mysterious. People may believe weird things because they want to, perhaps because it makes them feel good. Levy-Bruhl, not surprisingly, was fascinated by the report that the Boro Indians of South America considered themselves to be simultaneously human beings and red parrots.

One may attribute rationality to an action if it has a clear goal, and attribute rationality to a belief that is based on good evidence. The anthropologist Sir James George Frazer, who recorded many magical beliefs, argued that magical acts are the result of false beliefs, but that spells are genuinely thought to have the power of effecting the desired result. There is often the false belief that by imitating a described effect you can produce it, another example of representativeness. The real problem with magic for those who hold such beliefs is that the beliefs cannot be easily proved wrong. Experiments have shown that people with a strong belief in extrasensory perception, and those who scored high in magical ideation, did not show the expected right-side (left-brained) pattern of activity which is the normal way of assessing evidence. Contrary to paranormal explanations, one does not have to meditate, or have psychic instruction, in order, for example, to walk barefoot over glowing red coals – the ability to do so, walking not too slowly, is made possible by the physical nature of how heat is distributed among the coals. An essential feature of most beliefs in the paranormal is rationalisation, so when the belief is confounded by events, a kind of confabulation occurs. When the gods or witch doctors do not answer prayers, some reason is found for why they do not do so. If prayers for a sick child fail and the child dies, an explanation might be that the individual who prayed had committed some unforgivable sin.

Witchcraft and social structure are intimately related, so that small-scale communities such as those in southern Africa, with inadequate control of their surrounding environment, encourage people to seek personal causes for their misfortune. More generally, within unstable situations in southern African society, where wider environmental, economic or political changes have impacted on local communities, extreme stress and flash points of resentment can lead to witch killing. Where a local world was in flux, attacks on witches could thus provide an outlet for frustration, a means to secure access to material and scarce resources, or to redress unwelcome shifts in the balance of power. Beliefs in witchcraft were therefore exploited for a scapegoating process that involved a trajectory of violence against defenceless individuals. There was no common model, although there were often strong points of similarity between outbreaks. Often the victims were marginalised from their communities; for example, old women without kin, or herd boys geographically isolated from the village. There was frequently a pronounced gender or generational character, with the female and the old being common targets. Some of those targeted were, however, the conspicuously prosperous, as with black traders, whose success had aroused aggressive jealousy, and who were seen as trading core traditional values for economic success. A survey in southern Africa found that four fifths of those accused of witchcraft were related to their victims – most frequently a wife or mother-in-law was accused.

Human causation, as was discussed in relation to religion, is the cause in which we probably have the greatest confidence. All humans are exposed to other humans during their early years, and it would be surprising if they did not see humans as causal agents from a very early age. Children need to understand the actions of others, and try to predict how they will act, and what essentially causes their behaviour. Many myths invest stories with supernatural beings and strange happenings, and these can explain the origins of the world, of evil, and of heroes. Belief in monstrous humans has been widespread. Even Linnaeus, in his great classification of living things in the eighteenth century, included a species Homo monstrous on the basis that there existed weird relatives of Homo sapiens. For 2,000 years, there have been reports by travellers of people with gigantic heads, no heads, bird beaks and pig snouts. Herodotus, though sceptical of many reports, did describe people who had only one eye in the middle of their forehead. Later, St Augustine accepted the birth of monstrous humans, and mediaeval scholars argued that this was the result of the devil having perverted their souls. Columbus reported the existence of humans that were tailed or dog-headed, but by the nineteenth century, although there were occasionally such reports, they became increasingly rare.

Michael Shermer thinks it reasonable to explain the evolution of such myths in terms of a biologically determined module for making myths that helped our ancestors to survive. Shermer refers to the case of dragons and werewolves. Dragons are the most common of all mythological creatures, often being a hybrid like a lion's head on the body of a serpent. It could be, he suggests, that these stories helped instruct us about the nature of our environment, that lions and snakes were our enemies. Dogs, man's best friend, by contrast, do not feature in myths. Could some of these images and fears be encoded in our genes? Are we not programmed to fear animals like snakes and spiders? Recall the absence of phobias for electrical plugs, no matter how often parents emphasise their danger.

Aliens in flying saucers may be the modern technological equivalent of those ancient monsters. If aliens have been visiting the earth for thousands of years, then maybe at some point they gave us a helping hand. Erich Von Daniken made this suggestion in his enormously popular book *Chariots of the Gods*. Colossal prehistoric monuments such as the pyramids and Stonehenge seemed to him to be beyond the ability of savages and so must have been the work of extraterrestrials.

Von Daniken's work was pop archaeology at its worst, and critics easily shot holes in his theories, pointing out that Von Daniken was simply not giving prehistoric people any credit.

The modern age of UFO culture began in 1947 when an American pilot witnessed a formation of silver disks skipping through the air over Mount Rainier in Washington State. His report was widely circulated by the print and radio media, sparking interest in UFOs around the world. Just two weeks later, another event took place in New Mexico, an episode that has been at the root of more than half of today's UFO mythology. A rancher found the wreckage of something very unusual: scattered across a large area were bits and pieces of a thin, lightweight, silver material. It was extremely strong and flexible, with strange hieroglyph-like markings on some of the pieces. The rancher collected some of the debris and took it to the nearby Roswell Army Air Field. A press release written by the base's Public Information Officer was sent to the newspapers, claiming that the Air Corps had recovered a 'flying disk'. The release was retracted later the same day, when the Army stated that no flying disks had been found. Instead, what had been recovered was the radar reflector from a weather balloon.

After Roswell, flying saucers began appearing everywhere, including in Britain. Most attention has focused on the supposed alien bases, which consist of a US military installation on the surface and a vast alien complex underground. Alien abduction cases, revealed using hypnotic regression, have been widely publicised. The abductions that subjects described were nightmarish and often sexual. Victims found themselves helpless and unable to move while aliens appeared through their bedroom walls. Often sperm and ovary samples would be taken. The aliens seemed to be embarked on a genetic engineering scheme to create a hybrid human-alien race. Afterwards, women would find themselves pregnant, only to have the baby disappear from their wombs or to miscarry in the third month. One need only refer back to the hypnosis experiments in chapter seven to put all this in perspective. Indeed, those who believe in UFOs have been found to have a high level of hypnotic susceptibility. Suggestibility may also be a factor: seeing stories of UFOs on TV, or having read about them.

Belief in extraterrestrial 'beings' that can abduct, abuse and then return their victims has been widely reported. Shermer had contact with a group who believed that they had been abducted by aliens, and was surprised to find that none of them recalled being abducted immediately later, the experience, but that it was only years after that they recalled the abduction under hypnosis. Perhaps this is unsurprising when one realises that nearly one third of American psychotherapists accept that hypnosis facilitates recall of memories not just from this life, but also from some previous life.

There is a correlation between belief in paranormal phenomena and the ability to see patterns where there is little evidence for them, as in a random pattern of dots. There is also the correlation with the ability to fantasise, and the capacity to become strongly engrossed in the contents of an unusual experience. A major factor is a traumatic event in childhood, such as extreme loneliness or physical abuse. Fantasy could be an escape from severe stress. It could also provide a belief that there is a world over which children have control. Indeed, control may be a key in relation to paranormal beliefs. Paranormal beliefs can provide a cognitive framework within which life events are comprehensible and therefore may be mastered. Personal experience of any paranormal phenomena provides a very strong basis for believing in them. But there also seems to be a powerful urge to believe in such phenomena, and this, as I have suggested in relation to false beliefs, may be programmed in our brains. Belief in paranormal phenomena is very difficult to change.

Some 30 per cent of Americans believe in ghosts, 70 per cent in angels, and as many as one in ten has claimed to have seen, or had contact with, a ghost. These experiences include not just ghostly apparitions but also unusual smells, and the strong sense of someone or something being present. In addition, 25 per cent claim to have had a telepathic experience, and 11 per cent to have seen a flying saucer. Around 50 per cent of the population believe in ESP (extrasensory perception), and many US adults have themselves had experience of ESP.

A survey in the UK found that people with strong paranormal beliefs had a high level of religious interest, but were not integrated into religious institutions or events. They were seeking transcendence, but were compensating with their beliefs for not being in an organised religion. A survey in Reading found that those who believed in the paranormal based this on personal experience. Examples included dreams that came true, seeing a dead relative, and thinking of someone dying who was in fact dying. About 70 per cent believed in the paranormal. Their belief often came from friends and the media.

On that score, I have a close and highly intelligent artist friend who has reported to me that she has seen and communicated with ghosts on three separate occasions. She also says that they shared a common feature in not having feet. In each case they told her something about the place where she observed them that she did not know, but later confirmed. There is no question as to how real the experiences were for her. How different was her experience from those who have false beliefs that result in hallucinations? As described earlier, delusional beliefs are quite widely held across the general population. A specific study was made of the delusional component in the beliefs of those who believe in the paranormal. Paranormal believers scored higher on the delusion scale, and they also made more errors in a deductive reasoning task.

Many individuals claim to possess psychic ability. These include faith healers and those who can both predict the future and contact the dead. There is no reliable evidence to support their claims. A terrible case of such fraud occurred in the 1970s when the so-called Reverend Jim Jones attracted several hundreds to his cult by faking miracles, and eventually led them to mass suicide. So, what is the basis for what Richard Wiseman has termed psychic fraud? Deception can play an important role in determining what we believe. As early as 1887, a fake seance resulted in observer-invented events that did not actually occur.

Magicians and conjurors are, in a sense, the ultimate deceivers, for illusion is what they totally rely on. What makes an event magical is that it goes against our natural expectations about causes. The attraction of magic can be very clearly seen in children. The enormous success of the Harry Potter books illustrates this well: the children enter a world so unlike the one they live in, yet there are important similarities; magic must be learned, can be difficult, and one has to work at it. And one cannot but be struck by the popularity of movies about paranormal events, monsters and aliens. Even Hillary Clinton has been sympathetic to psychic counsellors by urging them to let her talk to historical figures who would understand her problems and advise her.

In 1997, the self-proclaimed Vedic philosopher Leahcim Remrehs offered psychic wisdom over a radio station to Chicago listeners. He told how scientific thinking, compared to New Age Enlightenment, restricted one's ability to perceive other dimensions, the future and the past. He invited callers to just give their date of birth and then ask a single question – this would allow him to tap into their cosmic vibrations. His success in telling the callers things about themselves and their relationships was very impressive and persuasive. But this psychic was really Michael Shermer (backwards) using his psychological skills.

This is a nice example of the Barnum effect. If you make some general statement, people will take it as referring accurately to something specifically about them. It is a technique used by clever mediums who not only claim to be able to contact the dead, but tell a person something important about their lives on first meeting. The extent to which the medium is self-deceiving is interesting. The so-called psychic works from the general; for example, 'I sense a tension in your relationship,' or 'You are wearing something very important to you.' As Shermer points out, around half of all Americans believe in astrology.

Richard Wiseman, who has done research into many apparently paranormal phenomena, investigated two locations that have a reputation for being haunted. The subjects had no prior knowledge as to which areas were classified as haunted, and which were not. They did indeed have more unusual experiences in the so-called haunted areas, but this does not implicate ghosts: the variance in the magnetic field and lighting levels are much more likely to be the cause. Given the ease with which abnormal perceptions can occur, we should not be too surprised by these findings.

Wiseman conducted an experiment to assess the reliability of testimony relating to seance phenomena. A total of twenty-five people attended three seances. A seance room had been prepared. All of the windows and doors in the room had been sealed and blacked out, and twenty-five chairs had been arranged in a large circle. Various objects – a book, a slate and a bell – had been treated with luminous paint and placed on a small table situated in the middle of the circle. Everyone was led into the darkened seance room and shown to a chair. An actor played the part of the medium. He first pointed out the presence of a small luminous ball, approximately five centimetres in diameter, that was suspended on a piece of rope from the ceiling. Next, he extinguished the lights and asked everyone to join hands. The medium first asked the participants to concentrate on trying to move the luminous ball, and then to try in the same way to move the objects on the table. After leaving the seance room, the participants completed a short questionnaire, which asked them about their experience of the seance. During the seances, the slate, bell, book and table remained stationary. Despite this, one quarter of participants reported movement of at least one of them.

Among the stratagems used by fake mediums is the insistence on working in total darkness, and having all those attending linking hands around the seance table, both of them techniques for preventing the detection of secret accomplices moving around the seance room.

An important feature for those who observe a psychic is the frame of mind in which they come to the event. Sceptics come believing they will be exposed to some sort of trickery, whereas believers in psychic powers expect a genuine display, and that they might even make contact with another world. These expectations do affect the observer's experience. Their later recall of what happened, and whether there really has been a demonstration of extrasensory perception, is greatly influenced by their beliefs – the believers recall psychic phenomena even when the demonstration has been unsuccessful and nothing actually moved, as just described.

Many individuals have reported experiencing extraordinary phenomena during darkroom seances. Eyewitnesses claim that objects have mysteriously moved, strange sounds have been produced or ghostly forms have appeared, and that these phenomena have occurred under conditions that render normal explanations practically impossible. Believers in the paranormal argue that the conditions commonly associated with a seance, such as darkness, anticipation and fear, may act as a catalyst to produce these phenomena.

It is only to be expected that 'psychics' are very keen to persuade their observers that they have genuine powers. Like many others, they may also be keen on personal fame as well as genuinely wishing to help people. What, one must ask, do they really believe? There is sometimes just pleasure in the success of the trickery, as with the Cottingley fairies. In 1917, two young girls living in Cottingley, a small village in Yorkshire, produced a number of photographs of fairies. The girls had been teased about their claims of having seen fairies nearby. One borrowed a camera from her father, and when the pictures were developed later that evening in her father's darkroom, they showed the girls with fairies. Finally, in 1983, one confessed that the pictures had been faked. The girls had drawn the fairies, cut them out and fastened them to the ground with hatpins. The two girls had no stake in the deception that could have brought them money. Ego and just plain fun are not thought to be sufficient, but in this case were.

Some performers follow the great magician Robert Houdini's advice in 1878 never to tell in advance what effect they will produce. However skilful the performer may be, and however complete his preparations for a given trick, it is still possible that some unforeseen accident may cause a failure. The only way to get out of such a difficulty is to finish the trick in another manner. But to be able to do this, the performer must have strictly complied with Houdini's important rule; and when working in groups, the advice is to create some chaos, move and talk fast.

Many people believe in telepathy and extrasensory perception. This belief is often linked to psi, an apparently scientific term that believers apply to an anomalous process of information or energy transfer, such as is supposed to be responsible for telepathy or other forms of extrasensory perception. There is no evidence whatsoever for it. Why people believe so fervently in extrasensory perception is that perhaps, for them, there is so much apparent evidence for it, and it opens up a whole new world outside modern physics. One's friends, for example, have strange experiences like premonitions and apparently extraordinary coincidences. There may be nothing more to it than coincidence, as persuasive scientific evidence is just not there. Moreover such processes are inexplicable in terms of what is now known about physics and biology. Yet, the public apart, many academics, including a few scientists, believe ESP exists. But most academic psychologists do not hold such beliefs, and argue that the evidence for such extraordinary events requires particularly reliable evidence.

An experimental method for studying ESP is the Ganzfield procedure, which tests for telepathic communication between a sender and a receiver. The receiver is placed in a reclining chair in a soundproofed room with translucent half ping-pong balls placed over the eyes, and headphones over the ears. A red floodlight is directed at the eyes and white noise played through the headphones. All this creates a total homogeneous field – the Ganzfield. The sender is placed in another isolated room and is randomly presented with a picture, photograph or videotape. The receiver provides a verbal report of the images he or she has perceived. Then the receiver is presented with several, say four, images and is asked which matches the images they received. Of the four images only one was 'sent' by the sender, so the chance of being correct, by chance alone, is one in four. There was in 1994 a report in a respected science journal, the *Psychological Bulletin*, that the receiver did slightly better than this, and claimed to show a paranormal transfer, but those who have examined the report closely have concluded that the necessary evidence is flawed.

There are also claims for psychokinetic phenomena where the mind alone, just thoughts, can affect the behaviour of objects both nearby and very far away. Science has no evidence whatsoever that the implied forces could exist. Such phenomena, it is claimed, could act to heal, bring rain, or raise a table. Poltergeists, non-human agents, must be assumed to use similar methods. The stories are legion, yet those with such powers have been very constrained in what they might have done – changed lead into gold, prevented ageing, and so on. And once again, why are such extraordinary powers so rarely used? Science provides no evidence whatsoever that the implied forces could exist.

Many psychics claim to be able to help the police solve serious crime. Psychics have been used in over one third of USA urban police departments and they have also been used in Europe. Wiseman and his colleagues compared the performance of two groups, one of psychic detectives, and the other a group of college students who were the control group. One of the psychic detectives was taken quite seriously by his local police force. Each group was shown three items that had been involved in one of three crimes: a bullet, a scarf and a shoe. The three crimes were, briefly: a soldier shot his wife and buried her, and she was identified by her shoe years later; a police officer was shot and the two men responsible were convicted on evidence that included the bullet; and an elderly lady was strangled by her scarf. All the subjects were asked to handle all three objects, and then filled in a questionnaire relating to the items, in which they had to say whether the statements about the crimes were true or false. None of the scores was statistically significantly different from random. However, the psychics, when told of the crimes involved, believed that they had in fact been successful – they had got some things right, like the scarf being involved in suffocation. But they ignored their false predictions.

Miraculous events are reported in many places, particularly

in religious texts. The evidence, to put it mildly, is rarely convincing, and usually relies on personal testimony. An interesting example of the reliability of this personal testimony is provided by what Wiseman refers to as the best-known secular miracle: the Indian rope trick. Accounts of this can be found in Buddhist and Hindu literature, and there is eyewitness testimony from the fourteenth century onwards. In the standard accounts, a magician in a field or market square throws up a rope, which remains rigid. A boy then climbs up the rope and vanishes. The magician calls the boy back, and when he does not appear, he himself climbs the rope and vanishes. Next, the boy's body, dismembered, falls to the ground, and the magician returns and restores the boy to normal health. Many individuals claim to have witnessed the trick, or at least some portion of it. Why do they believe they have seen such a 'miraculous' event? Wiseman surmised that whatever the reason, the reports would have become increasingly exaggerated over time from the reports of those who actually claimed to have seen it. And this is what he found: the average time that had passed since anyone reported seeing the boy climbing and disappearing was thirty years. There seems to be a real need for some to believe in such events.

The wish to believe in the paranormal can be rather strong. A stage magician performed fake psychic phenomena in front of two groups of university students. One group was told that he was a magician, while the other group was told he was a genuine psychic. When asked afterwards whether or not they believed he had genuine psychic powers, about two thirds of the students in both groups thought that he did. Even when the groups who were initially told that he was psychic were told that he was a fake, half still believed he had special psychic powers. Psychic powers are very seductive.

Here are two nice examples of how those with supposed

paranormal powers can work. The performer claims the magical power of being able to block the flow of blood in his arm at will, until the pulse actually stops; the reason is that he has a ball in his armpit which, when he presses down on it, stops the blood flow. Uri Geller has 'cured' many stopped watches simply by putting 'energy' into them by holding them in his hand. The reason is that in many cases a watch has stopped because it is jammed with dust and oil; holding it in his hand warms it up and frees it to work again.

The media play a key role in promoting supernatural beliefs – just think of the popularity of *The X-Files*. But can TV have the opposite effect by debunking the paranormal? A series in the UK, *James Randi: Psychic Investigator*, aimed to debunk psychic phenomena. A survey of attitudes, both before and after seeing the programme, found that about half believed that there was enough evidence to have serious discussion on the topic, and one quarter believed that ghosts were not imaginary. About one quarter believed in a supreme being with whom humans were not in contact. And about one third had had some sort of paranormal experience such as a premonition. The Randi programmes had virtually no effect on changing people's beliefs, but some TV critics condemned them as hypocritical and tasteless for even questioning the paranormal.

In the *Los Angeles Times* recently there was a long news report on a rainmaker. In Montana, it reported, there was a drought and they longed for this rainmaker, but things were so bad they did not have the money to employ him – around $10,000. He claimed a 99 per cent success rate and hundreds of satisfied customers. The report said he had brought rain to Montana in 2001. For $10,000 he brought rain to every little town he visited, failing only once in one small town. Deeply religious, he claimed to be able to shift the winds in fifteen minutes. His failure at the small town was, he later revealed, due to his problems with extraterrestrials there. Not one word in this two-page broadsheet report raised a single question about his claimed abilities.

With its promotion by the media, belief in the paranormal seems to be comforting to some, but is acutely irritating to others like myself. It suggests that there is a greater reality that we do not yet understand, and may never do. It opens up the possibility of life after death. It also opens up the possibility that we all have powers that we can develop to help us control our own bodies, and even the behaviour of other people and objects around us. But would people believe in ESP if it did not resonate with their own experiences, particularly coincidences? Recall that with just twenty-three people in a room, the chances of two having the same birthday are fifty-fifty.

It is now asserted by some that science itself is the modern superstition. There is even a plea for a more holistic view, closer to God, and the recovery of the sacred. One commentator has even written: 'The only hope it seems to me, lies in the re-enchantment of the world.' Science, as we have seen, has really only been in the public domain since the Renaissance, and continually goes against common sense, so most paranormal events are all too plausible, though rarely repeatable. We are quite casual about evaluating evidence that goes against beliefs we hold strongly. We will return to science after looking at health, where mysterious forces are again invoked.

CHAPTER TEN

Health

If a person is poorly, receives treatment to make him better,
and then gets better, then no power of reasoning known to medical
science can convince him that it may not have been the treatment
that restored his health.

Peter Medawar

Illness provides a central example of the need for humans to find a causal explanation for serious events, and also to find a response. It is intolerable not to know the cause of an illness or how to treat it. There will always be beliefs as to why someone got ill, and then better. It is probably one of the earliest causes ancient humans must have sought to explain. Not only does illness cause very severe stress, pain, and even death, it is also a serious evolutionary disadvantage to be ill. It is thus no surprise that beliefs about health and illness are so often linked to religious beliefs. Dealing with illness was probably one of the reasons for the origins of religion. Moreover, as with religion, mystical and supernatural causes are repeatedly invoked. Scientific medicine is a very recent subject, so most early beliefs about illness and health were very unreliable. As with religion, human-like agencies such as witches and evil spells put out by enemies were believed everywhere to be the causes of illnesses, and the one cause of which humans could be confident. Also, understanding illness was very difficult; even today most medical scientific explanations go against common sense.

There is an evolutionary model, based on archaeology and extant hunter-gatherer societies, suggesting what humans have believed about illness, and how it should be treated. In the first stage, hunter-gatherer groups managed their health collectively. The sick might be cared for, or have to leave the group and die, for the group could not spend too much of its resources on caring for the ill. When populations became settled, new diseases would become common, and new explanations sought. For example, epidemics could affect the young, killing many. In these circumstances, religious and 'medical' leaders would have to come up with new explanations and treatments. The new breed of healers would have included diviners, shamans, herbalists and priests. Labour in the agricultural communities could be very demanding, and there may have been a need for tonics as well as medicines. Some groups learned to exploit the active components of plants that were found, particularly in rainforests. This must have involved trial and error and disasters, as many are highly toxic. The local chief, or the healer, may well have been chosen because he was the healthiest in the group.

A most intimate connection exists between health, illness and religion. This relationship could have been critical in the origin of religious beliefs, dealing as some of them do with physical illnesses. Very early Chinese medicine was based on the belief that life was controlled by spirits and demons, and the idea that the ancestral spirits needed pacifying in order to avoid disease. The medicine of the Old Testament is entirely supernatural and religious, and there are no treatments. It is God who inflicts disease on people, as he did on the Egyptians. He puts to death but can also heal. 'Praise the Lord oh my soul; and forget not all his benefits – who forgives all your sins and heals all your diseases,' says the Book of Psalms. According to the ancient Greeks and their humours, mental illness came from the gods. In India, Vedic healing of disease summons demons from the patient and then puts them into an enemy. Buddhism emphasises personal faith. Early Christians believed that sickness, whether or not caused by sin, could be healed by prayer. Even in the West, between the years 200 and 1700, almost all mental disorders were understood in terms of demonic possession.

In Judaism, pain and suffering are seen as part of the fate of mankind, and can be punishment for sin. For Muslims, there is a similar view, and suffering can be thought of as a means of instruction on how to behave. Both religions instruct their followers to fight pain, as it is not part of God's paradise. In the second century, Clement encouraged Christians in pain to ask God for understanding as to why they were in pain, to remove it, and meanwhile to give them the strength to endure it. In the fourth century, St Augustine believed that God promised not health but rather eternal life in heaven, where there would be no pain or suffering. The presence of pain on earth was God's desire to heal swellings of pride, to provide punishment for sin, and to give a reminder of mortality. The idea that pain strengthens and purifies is also present in Buddhism and Confucianism. For the Buddhist, pain is a defining feature of human life, and is simply to be endured. The Confucians believe it is a trial, which is necessary for future well-being. Praying to God has been found to be quite common amongst those suffering severe pain. There is some evidence that Buddhist meditation, in particular, can be regarded as similar to prayer, and that it does help. One can see how valuable the possible force of prayer is to the more or less helpless individual suffering from severe pain.

Ideas about illness are very old, and there is no society that does not have such beliefs relating to causal explanations and cures, but there was virtually no chance of getting it right. The body of Sanskrit literature called the Veda goes back more than 3,000 years. It contains praise and worship of the gods as well as prayers for health. Several gods were thought to have particular healing powers, and diseases could be caused by evil spirits. Classical Indian medicine is Ayurvedic, and based on three bodily humours not unlike those of Galen in ancient Greece, and seven bodily constituents including blood and semen. Its medicines are mainly herbal. At the same time a range of diseases, especially children's, are held to be due to celestial demons who may be controlled by the planets, and who need to be pacified.

The earliest Chinese medical texts are over 2,000 years old. Chinese medical theories were founded on the belief that the human body represents a microcosm of the natural and social world. Thus, bodily processes are not only similar to those in nature, but inseparable from them. Health depends on internal harmony within the body, and with the environment. Illness can be the result of physical agents like cold and contagion, but also of ancestral displeasure. The stuff the world is made of is qi, which has been variously translated as 'air' and 'energy', but there is no equivalent Western entity or concept. Qi can stimulate change, and is essential for life; loss of qi leads to death. Acupuncture medicine claims that there are channels in the body carrying energy for each bodily organ, and qi energy channels are central to acupuncture: they are the sites where needles are inserted. There are lengthy texts describing the basis of acupuncture but science has done little to explain it over the more than 2,000 years that it has been practised. Crystal healing is based on the transmission of this energy, and faith healing also works by channelling it. Greek philosophers thought of energy as a fluid permeating a substance. Energy of some type still permeates current ideas in alternative medicine, as we shall see.

In early Mesopotamia, disease was diagnosed by examination of the liver of sacrificed animals, the liver being the seat of life. The hand of God was seen to be everywhere, as were spirits, sorcery and malice. Illness was also an omen. For the early Egyptians magic was key, and amulets and chants were widely used together with a variety of medicines. Early Arabic-Islamic medicine included a range of beliefs and practices based on animism, which invoked human-like forces. The two main forces were the jinn and the evil eye. Jinn were minor spirits, which could bring both good and bad fortune, but in relation to illness their influence was always negative, and they were held responsible for madness, diseases of children, and epidemic disease. The evil eye was summoned by special healers with spells, and it was believed to be responsible for accidental injury and an individual's illness, as distinct from a group's. Magical incantations were used against these malevolent forces, together with the wearing of charms ranging from rabbits' teeth to the sap of the acacia tree.

The development of scientific thinking in Greece took place against a background of traditional beliefs. Medicine illustrates this. While Hippocratic authors advocated some rationality in their approach to illness, there were several rather different groups offering help to the ill. For example, there were those who offered herbal remedies, and even blamed demons. But around 100 BC Lucretius could write: 'Life does not differ essentially from other matter; it is a product of moving actions which are individually dead.' He was saying that life and illness did not depend on some special life force and should be treated as a physical condition. There were indeed open public debates about the nature of the medical arts, in which the audience took an active role. Moreover, the spread of literacy helped.

Again, a key issue was the distinction between myths and logic. In the treatise *On the Sacred Disease* (410 BC) by Hippocrates there is a polemic against those who invoke demonic forces, and they are called 'charlatans' and 'magicians' and are accused of ignorance, deceit and fraudulence. *An Ancient*

Medicine, a central Hippocratic text, contained the conviction that using their method, the causes and cures of all illnesses would, in time, be discovered.

Thus, with the Greeks we have for the first time a completely different approach, almost a scientific one. Hippocratic medicine is specifically based on natural causes, and is totally independent of the supernatural. The received idea of, for example, a divine origin for epilepsy is totally unacceptable. 'Men regard its nature and cause as divine from ignorance and wonder, and this notion is kept up by their inability to comprehend it.' What a wonderful quote from *On the Sacred Disease*. There is not a hint in Hippocratic medicine of the gods being able to cure a disease. At last there was an appeal to reason. Humans were governed by the same laws as those that governed the physical world. From this came Galen's four humours – black bile, blood, phlegm and yellow bile – which were so dominant for the next 2,000 years of Western medicine, even though their claimed functions were wrong and many other ideas were unreliable, if imaginative, with, for example, an emphasis on bile and phlegm coursing through the body causing epilepsy.

Almost all non-Western theories of illness interpret illness as an injury imposed from outside, and assume some sort of aggression has taken place: illness does not occur by chance. Humans, or related forms, are usually seen as the key evil agents, which fits with the ancient idea that humans clearly can exert forces, both good and bad. Illness may also be due to some act, or a supernatural being, or may involve magic. Turning to current practices worldwide, there is significantly still only a small range of theories about physical illness. In a study of 1300 different cultures, 139 were examined in detail, and supernatural beliefs were found to predominate and to fall into three classes: mystical, in which illness is the automatic consequence of some act; animistic, in which the cause is some supernatural being; and magical, the cause being a malicious spell. Only four cultures believed in the Western concepts of viruses, bacteria and cells.

There are still attempts to try to develop a comprehensive understanding of the therapeutic process in shamanism, faith healing, herbalism, New Age healing, Ayurvedic and Chinese medicine, and Western biomedicine including psychotherapy. In some of these, the diagnostic process is intimately linked to that of therapy. Astrologers, for example, serve a healing function by giving an account of the patient's life in terms of the stars. Giving medicines to those who are ill is a magical rite in many cultures. The rites are combined with spells. The number of medicines used is very large. Representativeness, in the sense described earlier, namely – that like goes with like – can exert a significant influence on the medicines and treatment chosen. In relation to illness, it results in many people having assumed that the symptoms of a disease resemble its cause or cure. In ancient Chinese medicine, for example, people with problems with their eyes were fed ground bats, as they believed bats had exceptionally good vision. The Zande gave the ground-up skulls of the red bush monkey to those with epilepsy, as this monkey exhibits jerky movements.

The principle that like goes with like is found in an extreme form in homeopathy. Modern homeopathic medicine might be understood in these terms, for it is based on the 'law of similarity' or representativeness, and so one gives small doses of a substance to patients who suffer from symptoms caused by higher doses. In the late eighteenth century, the founder Samuel Hahnemann believed that every disease could be cured by giving the ill patient whatever substance could produce similar symptoms in a healthy person. However, some of the doses given in homeopathy are so diluted that on chemical evidence there are no molecules of the medicine in the water. But there are those who claim that the water can retain a memory of the medicine that it so diluted away. Nonsense, but it is hard to change beliefs. However, to its credit, the French National Academy has effectively called it mumbo jumbo and urged the government to stop subsidising homeopathic cures.

In central Australia at the beginning of the last century, the Aboriginal medicine-man would bend over his patient and suck vigorously at the affected part of the body, then spit out pieces of wood or bone. Pain was believed to be due to the presence of a foreign object, which had to be removed. Cause and effect beliefs are clear. In Indonesia, a variety of animals are eaten because they are believed to be health remedies. Bat hearts are eaten as a treatment for asthma. Other animal parts eaten as remedies for impotence include cobra blood, tiger penises and monkey brains. One woman swears that five years of cobra milk and powdered shark cartilage cured her intestinal cancer.

There are also beliefs that blame things such as taboos or the stars for illness. That the Zande in Africa believe in witchcraft, as Evans-Pritchard made clear, in no way indicates that they do not believe in physical causes and effects in much the same way that we do. Belief in death from natural causes and belief in death from witchcraft are not mutually exclusive; rather they supplement one another, the one explaining what the other cannot. The Zande accept a mystical explanation of the causes of misfortune, sickness and death, but turn to other explanations when social forces and laws require them. Thus, if a child becomes ill, it could be because the parents had broken a taboo, like having sex before the child was weaned, rather than witchcraft. Again, incest could result in leprosy in the offspring. Witchcraft can even be invoked to explain why the breach of a taboo has not been punished.

It is about health that the Zande most often consult their oracles. Even Zande in good health will consult an oracle at the beginning of each month. A negative response can lead to anxiety and depressed feelings, and they will try to undo the witchcraft responsible. The family of a sick relative will consult the oracle to find out who is bewitching the ill person, and a Zande who falls ill may ask a friend to consult a poison oracle on his behalf. The means by which the oracle determines who the witch is, and how to bring about a cure, can be complex. It can involve bringing a chicken wing, and giving poison to live chickens when naming someone who might have intended to injure the ill person. Another approach is to consult the rubbing board. Names are placed on the board, and the result depends on the final position of a piece of sliding wood. Once the witch is identified, there is a further complex social procedure for trying, in public, to persuade the witch to stop. In addition, every illness has special medicines for treating it.

In Papua New Guinea there are beliefs in powers and spirits that can respond to and influence human behaviour. Illness is explained by referring to things outside the body, rather than inside it. The causes of illness are considered in terms of agent and intention, and often the forces have a human-like origin, though food could be a cause of illness. A Western doctor working with Papua New Guineans found that his searching for signs and symptoms to diagnose the illness was irrelevant to them. What they wanted was to know who was the cause, the agent of evil intent. When a cricket jumped onto a man's painful leg at the same time as the grandchild of a man who had died from a bad leg passed by, he was sure spirits had entered his leg and made it more painful. But they accept causal explanations for a breast abscess if a baby dies, as the milk swells the breast and changes into pus. In many cases there are no explanations of the illness, just an acknowledgement that it existed. When a girl who suffered from fits drowned, the discussion shifted from the recognition of her illness to why she was allowed near the water alone, and so to possible sorcery. Explanations of the cause of an illness often relate to planting, harvesting or gardening: 'The spirits of the plants were annoyed at being injured.' For treatment, they would address the spirits in a loud voice.

Muti is the Zulu word for medicine. In its everyday form, its adherents – who include more than 80 per cent of the population of South Africa – use potions made from the country's indigenous herbs and plants to cure headaches and stomach ailments. More complex complaints call for animal parts such as crocodile fat, hawks' wings, monkeys' heads or dried puff adders. In a gruesome extension of the representativeness principle, some muti followers believe that human body parts can be used to heal them or imbue them with special powers. Human hands burned to ash and mixed into a paste are seen as a cure for strokes. Blood is given to boost vitality. Brains are used to impart political power and business success. Genitals, breasts and placentas are used for infertility and good luck, with the genitalia of young boys and virgin girls being especially highly prized as 'uncontaminated' by sexual activity, and therefore more pure and potent. Medicines to strengthen against the forces of the unseen were a matter of frequent resort. Unseen forces are still a matter of strong African belief.

Hardly any studies have investigated the success of all these mystical medical treatments, but one may assume they had some success, for why otherwise would they continue to be used? First, many illnesses get better on their own, given sufficient time, but, for the individual, as Peter Medawar put it very clearly in the quotation at the beginning of this chapter, people believe in any treatment that makes them better. Secondly, there is the placebo effect, which will be discussed later. Thirdly, it was virtually impossible not to have beliefs in the cause, and not to try to treat the illness. This was probably the primary driving force for such beliefs.

Witchcraft and the evil eye are not peculiar to so-called primitive societies as found, for example, in Africa, but were present up to the seventeenth century in educated Europe. Disease could be transferred and transplanted. It was believed a sick person should boil eggs in his own urine and bury them, then as the ants ate them, the disease would disappear. With whooping cough, the patient should stand on the beach at high tide, and, as the tide went out, so would the cough. Disease could also be transferred to the dead, so a patient should grasp a corpse at a burial. One should also note that Roman Catholicism promoted miracle cures and the healing powers of the sacraments and of saints.

The eighteenth century saw an attempt to make medicine more like Newtonian science. Harvey's discovery of the circulation of the blood was a powerful stimulus, but led initially to little of value for patients. The one eighteenth-century improvement in practical medicine that decisively saved lives was inoculation, and then vaccination, against smallpox. Germ theory only became established with Pasteur in the late nineteenth century. Bloodletting was believed in by doctors from Galen's times, and it was the subject of one of the first clinical trials, by Pierre Louis in Paris in the early nineteenth century. He found how ineffective it was. Patients had been unnecessarily bled, with serious ill effects, for nearly 2,000 years.

Mesmer's eighteenth century ideas about animal magnetism being a universal vital force were condemned in France as scientifically groundless, but it became accepted that those who committed suicide were mentally ill, and that it was not a mutiny against God. Influential nineteenth-century physicians believed that hysteria was the cause of bizarre episodes of apparent witchcraft that could be observed in the asylum, where patients believed they were in the clutches of diabolical powers, or experiencing angelic revelations. Charcot, the nineteenth-century French neurologist, used hypnosis to uncover hysteria, but failed to realise that the behaviours he observed were most likely due to the suggestions he made to his patients. Freud was much influenced by these explanations.

William Gladstone thought that everyone would be much healthier if they chewed each bite of food precisely thirty-two times – there are, after all, thirty-two teeth. And George Washington believed that a variety of illnesses could be cured by passing two three-inch metal rods over the afflicted area. Even today, there is enormous enthusiasm for remedies that claim to cure cancer and other serious illnesses, but for which there is just no reliable evidence.

Values can influence the definition of health and disease, as is illustrated by views on mental illness in America in the nineteenth century. Some physicians asserted with all their authority that women who enjoyed sexual intercourse, or who indulged in masturbation, were afflicted with various forms of mental illness as well as physical problems. There were also diseases that affected only blacks: one disorder was even named 'drapetomania', the overwhelming desire of a slave to run away.

The images of one's own body and the beliefs related to it can also affect beliefs in health and illness. All people share at least some intuitive sense of self, of existing apart from other humans. But, and this is crucial, how the different parts, like the mind and heart, are believed to relate to each other, and as well as to the rest of the society, are very variable. Different cultures attach particular significance to certain bodily organs. Among the French, Spanish and Portuguese, the liver is held responsible for many illnesses, while the English and Germans take a similar view of the bowels.

In Western biomedicine, the original separation of mind and

body, following Descartes, allows a scientific study of the body's function in health and disease. But among some non-Western cultures, holism and monism emphasise inclusiveness. There can be a conception of harmonious wholes in which everything in the universe must be understood as a single unit. One holistic concept is the Chinese yin-yang cosmology in which everything, the human body included, is in a state of dynamic equilibrium oscillating between opposites like male and female, hot and cold, yin and yang. In Japan, the Confucian heritage persists, and it is the family, not the individual, that is the fundamental unit. It is claimed that the Gahuku-Gama of New Guinea have no concept of personal self, as social identity and personal identity are indistinguishable. Other cultures, similarly, understand the individual only in terms of social relationships. The Cuna Indians of Panama believe they have eight selves, each associated with different parts of the body. These explanations are very different from theories of natural causation, which see illness as a normal activity gone wrong, like an infestation by, for example, worms. Nevertheless, what is striking about these very different beliefs is that they all attempt to provide a coherent and internally consistent set of ideas, even though the mechanisms are unknown. Cause and effect are clear, even if the mechanisms are not.

Religious beliefs can have an important impact on health, since gods are believed to exert powerful forces. Millions have made pilgrimages to holy places like Lourdes in search of healing, and places of healing exist for other religions, such as the holy temple in Jerusalem for Jews, and Mecca for Muslims. The Catholic Church has only recognised some sixty-four true healings among all those pilgrims to Lourdes, but religion could have an important psychological effect. Studies of patients have found that around 50 per cent (it varies in studies from 30 per cent to 80 per cent) of visits to the doctor for physical symptoms that have no obvious organic basis could be due to somatisation – the mind affecting bodily functions, often as a result of depression. This strong relationship between psychological stress and somatic symptoms suggests that religious beliefs could actually make things better by reducing stress. Religious beliefs and behaviours are inversely related to several of the risk factors for heart diseases. Lower blood pressure, for example, has a positive association with religious belief. The death rate from heart disease among Mormons is about 30 per cent lower than the general American rate. There are also studies showing that mental illness is lower in religious groups. One American doctor from Harvard was so impressed by studies showing that the incidence of our bodies are programmed to believe and are nourished by prayer, that he has become a believer in God.

For the Jewish followers of Lubawitch, the universe is in a state of disharmony, as Adam failed to pull together all the fragments at its explosive origin, reminiscent, curiously, of the Big Bang. Jews have a mission to restore harmony through ritual and charitable deeds. Do prophets actually derive their claim to authority from their ability to heal people? Jewish scriptures offer many examples of healing through divine intervention. The leader of the Lubawitch Hasidim movement every year receives hundreds of letters seeking advice about illness and misfortune. Often the advice offered in return relates to some impurity in the home that must be removed. In one case, a rabbi's heart condition was cured, so it is claimed, when he discovered that the scroll on his door – the mezuzah – had a Hebrew misspelling for the word heart in the command 'Thou shall love thy God with all thy heart.' In another case, a worn shawl – tallach – was claimed to be the cause, each of its thirty-two strands corresponding to its owner's pain-stricken teeth.

Contemporary beliefs about health are varied, and common current beliefs about health come close to paranormal beliefs.

One will still do almost anything to cure oneself of a serious, unpleasant, and possibly fatal illness. This is particularly true in our society where cancer, arthritis and ageing are still hard to treat effectively with orthodox medicine. Many general beliefs about illness have a rather simple explanation about the cause and the cure: one gets sick because of too little sleep, and gets better by sleeping more; or one gets sick because of germs, and better because of the medicine that kills them. Of considerable importance is the finding that those patients who believe they know the cause of the illness do better than those who do not. Moreover, we almost always make up a story as to why we get ill. And a story that some treatment has cured someone else makes it worth trying. This is fundamental in determining beliefs about curing illness. There is a universal urge to know the cause of the illness and its true nature. It is intolerable to be told by one's doctor that the nature of your illness is not known. Health is an area where causal beliefs play a key role, and many may be not that dissimilar to those of our ancient ancestors.

While many Americans attribute illness to heredity, diet, obesity, smoking, stress and alcohol, there is evidence for cultural diversity. Thus, African Americans and Latin Americans often include supernatural causes, such as God's punishment and the evil eye, and many middle-class educated Americans think about illness, its cause and cure, in ways that are alien to their doctors. These ideas help them make sense of their illness. This difference in the models of illness could account for some of the dissatisfaction patients feel with modern medicine.

One model embedded in popular literature depicts the human being struggling to maintain his or her integrity in the face of constant threats from nasty external influences such as poisons and germs, as well as threats that are generated internally such as depression and fear. The defence is made up of the nervous and immune systems, together with glands. According to this model, disease is the result of a failure to eat and exercise in accordance with nature's laws. The brain can be portrayed as a black box rather than being made up of complex nerve networks; this box releases chemicals and controls forces that can lead to the rupturing of heart muscles. Stress can affect the adrenal glands and lead to many diseases, including hypertension, peptic ulcer and asthma. Vitamins are also often perceived as providing a defence against a variety of illnesses. Americans' faith in dietary supplements is almost like a religious belief system. They believe that they have the right to deal with their health problems in the way that is most useful to them and also gives them hope about their health. It is hope that drives their beliefs in this area, and the more natural the additive, the better, for there is the belief that natural products are not harmful.

One model for colds, held by many sufferers, but for which there is no evidence, is that they arise from cold air or water coming in contact with vulnerable areas such as the top of the head, neck or feet. I was brought up to believe that I would get a fever if, after washing my hair, I went outside with it still wet. Another belief is that once a cold is acquired, it can migrate from the head to the nose and then down to the chest and bladder. From damp feet, a cold can travel up to cause a stomach chill. Avoidance of exposure to cold air or water is thus essential, and warmth and hot foods are sensible treatments. Fevers, by contrast, are believed to be caused by germs that come from other people, and are treated by starvation – starving the germs – and expulsion by fluids.

In Piaget's prelogical thinking stage – two to six years – children are unable to distance themselves from their environment. At this stage they see illness as a concrete phenomenon. 'How do people get colds?' 'From the sun.' 'How do people get measles?' 'From God.' More mature children think in terms of contagion: they can get a cold from someone else. At the next 'concrete-logical' stage, there is a clear differentiation between self and others. So getting a cold can be the result of your own behaviour, such as going outside in winter without a hat. The concept of a cold now involves ideas about sneezing and blocked noses, and hot air can push the cold back. After eleven years of age, physiological explanations begin, and there is talk of mucus and viruses. There is also now the idea that thoughts, particularly stressful ones, can negatively affect your heart.

Considering how much everyone cares about their health, it is surprising how little is known about what the public as a whole believes and understands about medicine, and particularly about mental health. What is even worse is that there are very few studies by the relevant authorities to find out what actually is understood. Interest has been focused on attitudes, not understanding. Science, including medical science, does go against common sense, but many of the misunderstandings are just due to ignorance and the unhelpful influence of the media. Recent surveys showed 96 per cent of the public had heard of diabetes, but they found that 48 per cent could not name any symptom, and only 4 per cent knew of thirst and polyuria in combination.

There is also evidence that patients have poor recall of what the doctor has told them, and a very poor appreciation of probability and risk: they often see 8.9 per 1,000 as a lower risk than 2.6 per 1,000. Moreover, if patients are told that 10 per cent will die with a particular surgical treatment or that 90 per cent will survive, they are less likely to accept the risk if told the former, and even doctors lean in that direction. There is, again, a very poor realisation of the importance of randomised clinical trials. My own questioning of friends about the difference between viruses and bacteria found that they know that one does not take an antibiotic for an illness caused by a virus, but have not a clue that a bacterium is a living cell, while a virus is not alive, but contains the code for its replication that can only be used when it infects a cell.

Many adults in the West do appreciate that inheritance, can, as a result of defective genes, result in illness, and that resemblance to a parent implies a likelihood of inheriting their illness. But there is also an overestimation of the shared inheritance in relation to parents, while the importance of siblings, aunts and uncles is neglected. More importantly, people have a very poor set of ideas about how genes can exert their effect – it is, in fact, a quite difficult science. Nevertheless, there are clear and correct beliefs that a variety of illnesses are inherited, including cancer, depression and heart disease. There is a widespread but mistaken view among children that one is most likely to take after the parent of the same sex, and thus inherit that parent's illness.

A very large number of people use alternative and unorthodox therapies when they are ill. There are some 50,000 practitioners of alternative medicine – or complementary medicine, which is a more favoured term – in the UK. About one third of the population make use of their services. Many GPs even provide complementary medicine treatments for their patients. Why are they so popular when most of them are at total variance with physics, chemistry and biology, as well as orthodox medical practice?

A distinction needs to be made among the almost thirty different disciplines that make up complementary or alternative medicine, especially with regard to evidence for their effectiveness. Thus, osteopathy and chiropractic, which involve manipulation, claim well-established benefits, even though the American Chiropractic Association attributes almost all disease to disturbances of the nervous system caused by the vertebrae of the spine – spine manipulation is thus a common treatment. There is also some evidence for herbal remedies and acupuncture being helpful. Crystal therapy, however, which claims that kinds of semi-precious stones have healing properties, defies physical principles. Again, iridology and reflexology, which claim to be able to diagnose various illnesses by examination of the iris and feet respectively, are without any scientific foundation or evidence. Aromatherapy is based on the healing properties of essential oils, which are supposed to represent the spirits of the plants they came from. In a way, homeopathy is related to holistic medicine. Holistic medicine rejects, or minimises, what is seen as the materialistic and reductionist basics of modern medicine, with its emphasis on cells and molecules and chemistry. Holists see psychological and even spiritual factors as being the cause or remedy. Emphasis is on the person as a whole, which could have advantages. Meditation and positive mental imagery are encouraged. There is the belief in homeopathy that the less of the substance taken, the greater the effect. This is contrary to everything we know about the natural world.

The concept of force has been used in determining many beliefs about the cause of illness, primarily in relation to some sort of energy. A belief in a special kind of energy has been common in medicine for centuries. After I had experienced Shiatsu, which is like acupuncture without needles, I was told that I had very low kidney energy. The idea that pounding and pressing bits of my body would detect such a localised quantity was irrational and absurd. Practitioners have taken a scientific term, 'energy', which refers to the capacity to do work, and used it in a way that seems to be totally inappropriate; but because the word is from science it gives it a spurious validity. 'Energy' medicine apparently accepts the existence of a subtle energy system within the body, holding that manipulating the body with force can cause physical, material changes to take place. The body is viewed as an integrated whole. There is no indication of how this energy is generated or what its nature is. Current believers in Chinese medicine claim that the pattern of distribution of the 'energy' qi can be understood in terms of yin and yang, which are polar opposites, but it can be very complex. Bodily functions are explained in similar terms: for example, yin relates to the heart, and yang to the intestine. All these concepts of energy bear no relation to the scientific concept, which is basically the capacity to do work, and is very well defined. This emphasises again the importance of beliefs in physical causes, which we acquire in infancy.

Mental health literacy among the public is of the greatest importance, but is sorely neglected. Almost every family in the land, as the Royal College of Psychiatrists points out, will at some time be affected by mental ill health. Literacy in this area could be defined as beliefs relating to recognising the different forms of mental illness, the risk factors and the causes, the role of self help, and when and from whom to seek professional help. There is virtually no research about such literacy in the UK. The general public does not apparently perceive psychiatrists as medical doctors, and what little reference they get in the press is usually negative. Of over 330 health-related articles in the UK press, only 47 were about psychiatry, and of these only 11 per cent were positive while 64 per cent were negative. Studies from Germany and Australia found that the public believes that mental health professionals can help with schizophrenia but not with depression, but patients who have received professional help for depression have more confidence in the value of medication.

The use of antidepressants is opposed by various groups and individuals for reasons that are hard to understand; there are those who wish to persuade the public both that they are dangerous and that they lead to dependence. Antidepressants have helped many thousands, and saved many lives. Such attitudes possibly reflect hostility to the pharmaceutical industry, as well as a belief that it is psychosocial rather than biological factors that are the causes of mental illness. Natural remedies, rather than medication, are very often favoured. Even the Nuffield Council on Bioethics, in their report on the genetic basis of mental illness, argued that one could not really identify genetics as a major cause, separate from environmental influences, and insisted that the patient be viewed as a whole. This ignores the overwhelming evidence for genes playing a key role in mental illness.

Non-medical causes of illness offered by psychiatric patients in a university hospital in the USA included 'God's will' and the hex or evil eye. Psychoanalysis and Freudian views of the unconscious present us with a related set of beliefs that I think fit most comfortably with paranormal beliefs. Energy appears again in the form of psychic energy, for which there is as little evidence as for qi. While the aim of Freud was to make psycho-analysis part of natural science, it has not turned out that way, and Freudian explanations seem to be much closer to beliefs related to witchcraft in the way they try to deal with mental ill-ness. Yet the concepts of repression, libido, and the Oedipus complex are repeatedly used by many people in the West as causal explanations for people's behaviour, both normal and abnormal. Is it not strange, and close to the paranormal, to believe that there are three mental processes in the brain that are almost like separate individuals – the ego, the superego, and the id – which interact with one another?

There are similarities between spirit possession and seances, and the contemporary psychotherapeutic clinic. The affected individual is no longer morally accountable for his actions, and a shaman or doctor names the 'spirit' responsible for the state. The resemblance between the psychoanalyst's ego, superego and id and certain Western African psychologies relating to soul, nature and lineage has been noted.

It has been suggested, and I support it strongly, that the popularity of Freudian theory is that it encourages people to do what comes naturally when they make causal explanations; that is, they use the representativeness heuristic. Dreams provide a good example. While there are many possible relationships between a dream and what has caused the dream, psychoanalytic theory assumes representativeness to explain all. So, in the interpretation of dreams by analysts, there is almost always a resemblance between the images in the dream and the supposed cause. Dreams about snakes relate to penis problems, those about policemen to authoritarian fathers.

The true advantage of the psychoanalytic mode of thought is that it is accessible to almost anyone, and there is nothing for which an explanation cannot be obtained, making it very seductive. The concepts are so poorly defined that they can explain any behaviour. And those three core concepts – id, ego and superego – seem to be more complex than anything that they claim to be able to explain. For example, the classic case of Anna O. was accounted for by the trauma of her seeing a dog drink from a friend's glass, but how does one decide it is a trauma, and is there anything that may not thus at some stage become traumatic?

Belief in a treatment can itself have important consequences. One incident in the Second World War illustrates this. Henry Beecher, an American anaesthetist working at the front line, ran out of morphine. A nurse injected a soldier with severe injuries with salt water, and the patient settled down and felt very little pain, just as if he had been given morphine. This is the origin of the concept of the placebo effect. From further work in 1955, Beecher claimed that placebos were capable of producing gross physical change. Many of the studies were flawed, as there was no proper control group (those who received no treatment at all and who improved). Indeed there are those who have examined the trials carefully and concluded that placebo is no more than a myth. Yet there is very good and reliable evidence that all sorts of pain – headaches, post-operative pain and even sore knees – can be relieved with a sugar pill placebo.

In one trial with ultrasound for post-operative pain following tooth extraction, neither doctors nor patients knew when the machine was on. Compared with those who had no treatment, all those treated with 'ultrasound' did better. The very presence of the machine was important, for the placebo effect is intimately associated with the desires and expectations of the patients. There is also positive evidence for a placebo effect with angina: some patients had their arteries exposed but not treated, and yet they improved as much as those who received actual surgical intervention. And in the case of drug treatment for depression, there is also strong evidence for a placebo effect. It may be that placebos are only effective in certain disorders.

One has, however, to be quite careful with trials that claim a positive effect for a placebo, since patients can get better on their own. There is no good evidence that placebos work in relation to cancer or schizophrenia, but there is good evidence for the positive effect of placebos on pain, inflammation and depression. Just why belief in a treatment – and belief is essential for the placebo effect – should have this effect remains unclear. But for the placebo to work, one must believe that one has been given a powerful medical treatment. In an African tribe you need to believe in the shaman, not a doctor.

What is the nature of the belief that can result in a placebo response? Herein lies a problem, because people can have contradictory beliefs, and may act in ways that are inconsistent with their beliefs. Part of the problem is that one cannot view a belief directly. One way to construct what someone believes is to do so on the basis of their behaviour, including what they say. So what beliefs give rise to the placebo effect? One important factor is that some medical intervention for the patient's illness appears to be actually taking place. It is necessary to believe that one is receiving a powerful treatment, the nature of which can vary greatly from culture to culture. One must also believe that it will work. Further support for this is that a placebo given by injection is more effective than, for example, pills. Again, if patients are first given a painkiller, and then given a placebo, the effect of the placebo is greater since the painkiller increased the belief that the treatment would work. However, this may reflect a type of learning and conditioning.

Unintentional communication between doctor and patient can influence the placebo response. In one study, patients who had undergone tooth extraction were given a pain reliever, a drug that increased the pain, or a placebo, just saline. They were, however, divided into two groups: in one group all three treatments were randomly assigned, while in the other the pain reliever was not included. While the doctors did not know who was getting a particular treatment, they knew which group any particular patient was in. All the patients in the group getting the pain reliever had much greater pain relief than those in the other. The doctor's expectations that they might be getting the pain reliever affected them in some subtle way.

The other, and unpleasant, side to the placebo effect is termed the nocebo. Nocebo involves getting ill or having unpleasant symptoms because of the expectation that this will happen. For example, 80 per cent of hospital patients given sugar as an emetic vomited, and asthmatics have had an attack caused by neutral inhalant which they were told would cause one, and cured by the same inhalant when told that it would help them. Medical students' disease is well known: many students begin to get the symptoms of the disease they are studying. It is also the case that depressed patients have a greater probability of heart disease because, perhaps, of their negative expectations with respect to their health. The most dramatic example is voodoo death, which has been reported in diverse cultures in Africa, South America and Australia. Its success depends upon the victim knowing the spell or ritual curse has been cast. Somatisation, how the beliefs held can affect the body, could be a related phenomenon.

The most reliable way to assess any medical treatment is by randomised clinical trials. The best and most satisfactory technique for doing this is called double-blind testing, in which the treatment is given to some patients and not to others. The patients are assigned randomly by the controller of the trials to one or other treatment, and neither the patients nor the doctors know who has been given which treatment; both, in this sense, are blind. This anonymity is essential because if either group knows what is going on, this can affect the outcome in all sorts of subtle ways. Not doing a proper clinical trial can lead to very misleading results; neither anecdote nor correlation will do. Consider, for example, my very reliable treatment for flu. Each morning when you are ill you get out of bed and sing 'God Save the Queen'. I can guarantee you that in three weeks you will be fine again. Recall, always, Peter Medawar's comment.

As with religion, we can see how deep-seated mystical forces are used again and again to account for serious events that affect our lives, like illness. We need explanations as to the causes, and in many cases these have a human character, or a mystical one, not unlike those associated with religious beliefs. Representativeness also plays a role. It is hard to resist thinking that many of the beliefs about illness reflect some primitive programming of our minds by our genes. Many beliefs about health are scientifically invalid. Just how hard it is for the people to form reliable beliefs about health is shown by the fact that at least 20 per cent of the public were persuaded by the claims of a solitary doctor that MMR causes autism. But before looking at science we turn to moral issues.

CHAPTER ELEVEN

Moral

> There are three forces, the only three forces capable
> of conquering and enslaving forever the conscience of these weak
> rebels in the interests of their own happiness. They are:
> the miracle, the mystery and authority.
>
> F. Dostoyevsky, *The Brothers Karamazov*

The moral and ethical beliefs that determine people's actions in their social interactions are of great importance. They are not causal beliefs as such, but determine what causal action an individual may take, particularly in relation to political and social issues. The consequences can play a central role in society, and can have devastating effects, as they are often the beliefs of those with power, exerted on others who do not share those beliefs. At a day-to-day level, there are very strong beliefs about such topics as the environment, nuclear power, genetic modification of humans and plants, and the rights of embryos. Much more important are the horrific examples of religious wars, racism, and the Holocaust. People in different countries become mutually trapped by their beliefs about each other, and such beliefs can be reinforced by government misinformation.

Many millions of people have been killed in war because of their beliefs and those of their leaders and enemies. The number killed in the Soviet Union since 1917 is around 60 million. What are the beliefs responsible for these acts? It has been argued that Stalin believed in all his terrible atrocities being for the best in the long run. How, also, can we understand the beliefs of people like suicide bombers? A serious problem, about which volumes will be written, relates to the belief of powerful Western politicians that Saddam Hussein had weapons of mass destruction ready for immediate use. By contrast, there are beliefs in many societies that lead to cooperation and altruism, as we will see.

There is evidence that there are common unconscious beliefs and preferences, and that these are involved in creating stereotypes and prejudice, including social prejudice. On this view, all humans are implicated in prejudicial attitudes. Judgements of individuals can be shifted in a more negative direction when they follow the activation of a belief about another person's social group. Forced beliefs are beliefs manufactured to support other beliefs that are poorly supported by evidence. A current forced belief is the denial that males and females are different in ways other than those related to reproduction, though research shows clearly that they have different and complementary mental and physical capabilities selected during evolution. People rationalise their actions in order to reduce discrepancies in their belief systems. Prior commitment to a belief system and personal sacrifice put up barriers that block the resolution of any conflicting evidence. Both US soldiers that fought in Vietnam, and those who avoided the draft, were committed to believing that they had made the correct decision, and that the others were cowardly, or would not face up to the truth.

Ethical and moral beliefs are clearly not held in isolation, but are part of a system in which various beliefs differ in the conviction with which they are held, and where some will be in conflict with others. For beliefs strongly held, it is often the case that there is virtually no evidence that will make someone give them up, and any evidence will be negated by various arguments not too unlike confabulation. In addition, particularly in relation to politics, the belief system is based on authority, which requires that the general system be adjusted to fit with that of the rulers. Self-deception, perhaps forced by fear, is common, as is a change in beliefs to follow the party line.

In a number of animal societies help is given to close relatives since it helps conserve the genes they share. Human societies are unique amongst animals in that many societies are based on cooperation between genetically unrelated individuals. There is evidence from sociobiology that some aspects of personal commitment and altruism are coded for by our genes by controlling the development of the brain so that these behaviours are instinctive. This is particularly clear in relation to our helping those who share our genes. But altruism, generosity and forgiveness in relation to those who are not close relatives may also be programmed in our brains.

When individuals believe that the other members of the group will be cooperative, they too will cooperate, but will be very disappointed when this turns out not to be the case. This could lead to breakdown of cooperation in the group as a whole. Even when there are a large number of cooperators in the group, studies have found that a small number of selfish individuals can result in zero cooperation throughout the group. To maintain cooperation, it is essential that most members believe that almost all of the other members are cooperative. This belief will be supported if there are known to be punishments for selfish individuals. There is also evidence that a considerable part of human altruism is driven by beliefs concerning what effect it will have on reputation. There is no doubt that our biology has given us the capacity for love and kindness, and so it should come as no surprise that it has also given us spite and revenge.

A person who makes no emotional commitment can expect none in return. This commitment could be a key reason for people to value religion. While it can be difficult to create committed relationships one by one, if you are a member of a group which vows to help its members, then this could help create a community of altruistic believers. But authority can pervert friendly cooperation.

The fundamental role authority can play in a social situation is illustrated by Stanley Milgram's study in the USA in the 1960s. Those who had volunteered to be subjects of the study were told that they were helping in a learning experiment. When the person learning, who was in a different room and could not be seen, gave a wrong answer, the subject was ordered to press a switch that gave the learner an electric shock. There were switches with increasing voltages marked on them, and the learner was given increasingly strong shocks. With the increasing strengths the learner complained, and eventually screamed, but the subject was told to continue right up to the switch labelled 'danger'. More than half of the subjects accepted authority right up to this level and gave what they must have believed were very dangerous shocks to an innocent learner. What is the adaptive advantage to the individual of accepting such orders, and to what extent has this been influenced by evolution? To what extent is it genetically determined? Perhaps obedience to the leader in early groups, and belief in their views and authority, was essential for the survival of the group.

Marx believed that he had discovered scientific laws governing human behaviour. In the early Soviet period there was the belief that it was possible to build a socialist society in which people would no longer exploit each other and where everyone would flourish. There was faith that it would succeed, and the Communists gave it the authority of applied science. Marxism stressed that people become deformed by the economic and social system and that it was necessary to make a Communist out of Capitalist material. Political action, for Marx, would be based on his theories. Morality was, for him, merely a disguise to protect class interests, and there was an almost religious commitment to Marxist beliefs.

This was perhaps not all that different from René Descartes' view of the authority of God:

> But above all we must impress on our memory the overriding rule that whatever God has revealed to us must be accepted as more certain than anything else. And though the light of reason may, with the utmost clarity and evidence, appear to suggest something different, we must still put our entire faith in divine authority, rather than in our own judgement.

Related beliefs were at the core of Stalinism, according to Jonathan Glover. Again, it was Marx who argued that the practical results of beliefs were the criteria for truth, and that practical results were what mattered, not whether the beliefs were true. The pressures exerted by the state were very severe; even inside families parents expressed no doubts to their children.

In China, too, many believed that a planned socialist economy might eliminate poverty. In 1957 Mao said, and presumably believed, that in the next fifteen years the Soviet Union would overtake the USA in steel production, and China would overtake Britain. Mao was very stubborn about refusing to hear what he did not want to hear. Ignoring advice on the dangers, he swam in the Yangtze and then proclaimed: 'There is nothing you cannot do if you are serious about it . . .'

As Glover puts it, the belief system was central to Nazism. The pressures to obey worked because the beliefs were accepted, even though they were in conflict with previously held moral beliefs. George Orwell suggested that Hitler's appeal came from his recognition that people want more than comfort and security; they want in addition struggle, sacrifice, drums and flags. Hitler was also helped by a culture that had great respect for authority, and lacked a tradition of criticism. This provides further evidence that there may be an innate disposition to conform to political beliefs and to obey authority.

Colonisation of the New World saw extreme expressions of racism – death by starvation and disease of the local peoples – almost amounting to practices of ethnic cleansing and genocide. The subjugation of the native peoples of the New World was legally sanctioned. In the fifteenth century, two papal bulls set the stage for European domination of the New World and Africa. Romanus Pontifex, issued by Pope Nicholas V to King Alfonso V of Portugal in 1452, declared war against all non-Christians throughout the world, and specifically sanctioned and promoted the conquest, colonisation and exploitation of non-Christian nations and their territories. Another bull by Pope Alexander VI in 1493 to the King and Queen of Spain, following the voyage of Christopher Columbus to the island he called Hispaniola, officially established Christian dominion over the New World. It called for the subjugation of the native inhabitants and their territories, and divided all newly discovered or yet-to-be discovered lands into two, giving Spain rights of conquest and dominion over one side of the globe, and Portugal over the other. Such political laws and convictions underpinned racism, South Africa being a clear example.

Racism is founded on the belief in one group's racial superiority over another group who have different physical or social characteristics. It could have a biological basis in helping to promote group identity. Racism encompasses the beliefs, attitudes, behaviours and practices that define people, and is based on the classification of human beings into distinguishable groups according to immutable physical characteristics, for example, skin colour, hair texture, eye shape. Racism uses the inflexible assumption that group differences are biologically determined and therefore inherently unchangeable. In South Africa the Bible was used by the Dutch Reformed Church to justify Apartheid, and it is to their credit that they later apologised and declared Apartheid a sin. Whether they all believed in their retraction is another matter.

Racial prejudice is, unfortunately, well illustrated with reference to the Jews. Beliefs that Jews are evil causal agents have a long history, going back to the crucifixion. During the Crusades the Jews were a main target, since they had murdered Jesus, were rich, and showed contempt for gentile ways. The Black Death in the middle of the fourteenth century killed some 25 million people. The rich blamed the poor and astrologers looked to the stars. The church explained it as a divine punishment for the world's sins. Then it was blamed on the Jews, who were said to have deliberately poisoned the wells. In the nineteenth century, anti-Semitism was based on religious convictions together with hope for material gain by dispossessing the Jews. There were fanatics, near psychotics, for whom Jewish crimes became the central focus of their lives.

The Nazi belief in the evil of Jews and the need for their extermination was fuelled by the national sense of humiliation following the First World War, and a hope of national renewal. Jews became the scapegoat for losing that war. For Hitler, not only were the Jews Marxists who misled and exploited the workers, but they were also out to control the world. The Jew was also a racial polluter. Hitler claimed that Jews did not join the army and ruined the economy. He also believed in racial purity, a view bound up with a myth about true German origins in the woods and forests. Scientific claims were made for the validity of such beliefs that related to racial hygiene, which aimed to protect the gene pool of the race. There was compulsory sterilisation for numerous illnesses including schizophrenia, hereditary blindness and severe alcoholism; and aliens, such as Jews, needed to be eliminated.

The scientific justification for racial purity had its origins in Britain at the end of the nineteenth century, with Francis

Galton and his eugenics programme, which he defined as the science of improving the human stock by promoting the reproduction of a better human race. Many scientists had beliefs that supported eugenics, such as that prostitution and poverty were inherited characteristics. The Nazi programme was supported by distinguished scientists such as Konrad Lorenz, who won a Nobel prize for his work on animal behaviour and wrote: 'There is a certain similarity between the measures which need to be taken when we draw a broad biological analogy between bodies and malignant tumours, on the one hand, and a nation and individuals within it who have become asocial because of their defective constitution.'

The use of biology to support belief came from a doctor, who said that the Jew was a gangrenous appendage in the body of mankind. For non-scientists such beliefs, apparently based on science, could be very persuasive. Again, the philosopher Nietzsche said that: 'The extinction of many types of people is just as desirable as any form of reproduction.' But even more powerful than pseudo-scientific reasoning was obedience forced by fear of authority. There was a belief that it was a moral duty to obey. Adolf Eichmann, who controlled the deportation of Jews to extermination camps, used this as a justification for his behaviour.

The beliefs associated with anti-Semitism and the murder of Jews in large numbers provide a valuable example of how such beliefs are formed, and how they affect behaviour. My source is Browning's *Ordinary Men*, in which he examines the behaviour of the men in a police battalion that killed over 30,000 Jews, mainly in Poland, in 1942. He wants to understand how they could do this, and my understanding of his interpretation is as follows. The men in the battalion included some of middle age, and of these some, including a commander, found the shooting of Jews very upsetting. Some even managed to get out of such duties. The reasons Browning offers for the behaviour of those who did the killing is that it was due to early indoctrination as to the evil nature of Jews and their danger to the German state, obedience to authority, and the wish to conform with the behaviour of their colleagues. Nazi supporters felt their lives were being given a special meaning by becoming part of a shared project; it was tribalism, not the rights of man, that was a dominant feature.

An important ethical issue relates to so-called brainwashing, a technique for changing someone's beliefs by physical or mental force. But does brainwashing really exist, even though it is widely believed in by the public? Modern brainwashing theories misrepresent earlier academic work on coercive processes developed in Russia, China and Korea. Those techniques were generally rather ineffective, and the efficiency of brainwashing techniques used by the communists during the Korean War, and subsequently by the CIA, is problematic. Simple group pressures and emotion-laden tactics are revealed as more effective than the tactics used in physically coercive situations. But there is insufficient research to permit informed responsible scholars to reach a consensus on the nature and effects of non-physical coercion.

It is quite often claimed, however, that modern cults effectively use such methods to entrap new members. There are some 5,000 economic, political and religious groups operating in the United States alone, with 2.5 million members. Over the last ten years, cults, it is claimed, have used tactics of coercive mind control to negatively impact an estimated 20 million victims. Satanic cults are claimed to exist on a local, state, national and international level, and there are weird claims that Satanists ritually kill tens of thousands of infants every year in the USA. Other infants and children within the cult are believed to be programmed to respond as robots without any degree of self-will. Victims can allegedly be triggered at a later date by sounds, words, images or colours to mindlessly perform prearranged acts in support of the Satanic cult. There is no real evidence for any of this.

Cult methods of recruiting, indoctrinating and influencing members are, in fact, not exotic forms of mind control, but the same mundane tactics of social influence practised daily by all compliant professionals and societal agents of influence, applied more intensely. Cult leaders offer simple solutions to the increasingly complex world problems we all face. They offer a simple path to happiness, to success, or to salvation by following simple rules, simple group regimentation, and simple total lifestyle. Ultimately, each new member contributes to the power of the leader by trading his or her freedom for the illusions of security and glory that group membership holds out.

An investigation showed that out of 1,000 people persuaded by the Moonies Unification Church to attend one of their overnight programmes in 1979, 90 per cent had no further involvement. Only 8 per cent joined for more than one week. Another indicator of the non-existence of true mind control, or brainwashing, is the high turnover rate of members: there is a 50 per cent attrition rate during the members: 'first two years. The opinions of former members who have left on their own do not believe that they were ever the victims of mind control'. Cult mind control is not different in kind from everyday varieties of attempts at persuasion, but differs in its greater intensity, persistence, duration and scope.

It is very difficult to understand the beliefs of criminals, particularly psychopathic murderers. It seems that they have diminished negative reactions to violence, and may lack normal social beliefs, for reasons that are poorly understood. And how does one understand suicide bombers or those who give up their lives so easily? The Japanese kamikaze suicide bombers in the last world war believed it was their duty to die if neces-

sary for their Emperor, a godlike figure. They also believed it
was the only way effectively to deal with the enemy. Are such
beliefs that different from those of the Allied soldiers, for example
on D-day in 1944 when they stormed the beaches and thousands died?

Terrorism occurs worldwide. In Sri Lanka the Tamil Tigers
had many suicide missions that had no religious basis except
cultish devotion to their leader. The blunt fact is that suicide
attacks are often based on the advantages they bring to the
group who commission them: it is a good way to kill people.
It is also helped by the belief of those who die that they will be
rewarded after death. There is only one verse in the Koran
that contains a phrase related to suicide: 'O you who believe!
Do not consume your wealth in the wrong way – rather
through trade mutually agreed to, and do not kill yourselves.
Surely God is Merciful toward you.' (4:29) The prophetic tradition clearly prohibits suicide. Ultimately, it is God, not
humans, who has authority over the span of every person's
life. There are some Muslims, most notably during the last
several decades, who have engaged in suicidal military missions. These extremists cite passages in the Koran that
promise paradise to those who die 'struggling in the way of
God' (2:154). They see what they are doing as active armed
struggle in defence of Islam. Their death is thus viewed as
martyrdom, not as suicide. Suicide bombers in London have
led to further attempts being made to understand their
beliefs. Interviews with young Muslims who share the views of
the bombers suggest that there is a belief that it will purify
their soul and that Allah will bless their action, and they will
meet their creator in an afterlife. The influence and support of
like-minded friends is important.

There is, however, a mystical dimension to Osama bin Laden,
who envisions Saladin, the Islamic hero of the Crusades, coming

out of the clouds. For bin Laden those who do not share his religious beliefs are bad people, and must be destroyed. For Muslims in general, Jihad means 'striving or struggling in the way of God' and doing the will of God. It means that because of human pride, selfishness and sinfulness, people of faith must constantly wrestle with themselves and strive to do what is right and good. The 'lesser' Jihad involves the outward defence of Islam. Muslims should be prepared to defend Islam, by military means if necessary. Meeting death while carrying out a Jihad incurred in Allah's name earns eternal bliss: '. . . they find their sustenance in the presence of the Lord'. While the vast majority of Muslims clearly reject the violent extremism of terrorists as manifested on September 11th, some religiously inspired and politically motivated adherents justify their behaviour in the context of a holy war. Their acts kill, but is this really different from those of any soldier who engages in battle? One difference is that they are killing civilians. But what did Allied bombers of cities do in the last world war?

A rather different set of ethical beliefs relates to the very nature of life. Beliefs about the nature of life have been an issue throughout history. When, it has been asked, does life begin? And where does it come from? For the ancient Egyptians, the sun god was the creator of the germ in women, and the seed in men. For the Greeks also, the origin of human life was an important issue. Plato thought the foetus a living creature as soon as it moved. Aristotle was sure the foetus had a soul, which entered it at some time, but at different times for males and females – later for the latter. There are twelfth-century drawings of a soul entering the embryo. St Augustine put animation – entry of the soul – at forty days for men and eighty for women, but then later set forty days for both sexes, and after this time the embryo could be baptised. In 1745 Cangiamila wrote *Embryologica Sacra* and advocated caesarean section to ensure baptism. Doctors at the Sorbonne in

Paris even recommended intrauterine baptism by means of a syringe. 'How many aborted foetuses should not have been received by the lavatory through the ignorance of their mothers when their souls, if not cheated of baptism could have gazed on God for eternity, and should more properly been given burial.' This is the origin of the idea that the embryo is human from the moment of conception. Even among early scientists, there was the view of the soul as the organising force of the embryo, and in the nineteenth century no scientist believed that the woman contributed anything important to the embryo other than nutrition: the sperm was the organiser.

With the ancient Greeks we have an association of life with the soul that is present even today, and is a belief that underpins vitalism, the idea that there is a mystical life property that cannot be accounted for by physics, chemistry, or any other science. There are a number of other beliefs about biology that apparently have a significant ethical content, particularly those relating to the early embryo. From the moment that the egg has been fertilised by the sperm, the status of the early embryo is that of a human being according to most Christian religions, and this is particularly clearly stated by the Catholic Church. This belief is of great importance: it could prevent assisted reproduction in which the egg is fertilised in a dish and then transferred to the mother, as this method leads to the destruction of many human embryos. Similarly, it would prevent prenatal diagnosis for genetic illnesses. Such beliefs could also prevent the termination of a pregnancy, even if the child would be severely handicapped.

There is, in fact, no biological basis for such beliefs, if only because the fertilised egg can develop into more than one person, namely, it can give rise to twins. I would argue that the embryo is only a person when the baby can survive outside the womb with minimal technical support. The divine basis for these religious convictions is far from clear, as it is of relatively recent origin. For hundreds of years, as mentioned above, the soul was thought to only enter the embryo around forty days after conception.

Our treatment of the environment involves a number of very divergent beliefs. For example, there are those totally committed to organic farming, though the evidence of its advantages is controversial. For some of those who strongly believe in organic farming, the idea that its achievements should be scientifically tested is unacceptable, because organic farming is holistic and cannot be studied by reductionist science. Some 90 per cent of the UK public would not eat GM foods, largely as a result of pressure groups who oppose them, and, of course, the pervasive influence of the press. Genetic modification is feared, and the influence of Mary Shelley's Frankenstein is pervasive. She clearly touched upon a basic fear of the unnatural. New technology and scientific discoveries are met with apprehension. Are scientists now all that different, in the public view, from witches?

More about science next, where it should become clear that reliable scientific beliefs have no intrinsic ethical or moral content: they refer to how the world is. It is only with technology – making or changing things – that ethical issues arise.

CHAPTER TWELVE

Science

A theory that fits all the facts is bound to be wrong,
as some of the facts will be wrong.
Francis Crick

Scientific beliefs are among the most important beliefs in human history. Scientific beliefs are special, and different from any other kind of thinking. Scientific thinking is not programmed in our brains, and only one society invented scientific beliefs. Logic and evidence dominate, and therefore many key beliefs in religion are in some conflict with science. Scientific thinking would not necessarily deny the existence of God, but would claim there is no good evidence to support such a belief. Moreover, beliefs in miracles and the paranormal go against some of the basic tenets of physics, chemistry and biology.

Science is the best way of understanding how the world works, the actual nature of causes. It is a communal enterprise, with the individuals contributing to a common body of knowledge, although the topics range from physics to the mind. There is no one scientific method other than to be internally consistent and to have explanations that fit with the real world. For any set of observations, there is only one correct scientific causal explanation.

There are many styles of doing science, from theory, to experiment, to careful observation. Does science provide beliefs that are fundamentally true? In general the answer is yes, though evidence can always make beliefs in those truths subject to change. It is implausible in the very extreme that DNA does not code for proteins, or that water is not H_2O. One can be very sure that over 90 per cent of chemistry is correct, and always will be, and so is the vast majority of physics. If the history of science were to be rerun, its history would be different, but the conclusions would be the same. The individual scientist, unlike the artist in the arts, is ultimately irrelevant, and the scientific genius merely speeds up progress. Science is, with rare exceptions, independent of cultural beliefs.

Why do scientists believe that science provides fundamental explanations for how the world works, particularly the nature of causes? An essential feature of science is that scientists contribute to a common body of knowledge, which is scrutinised by other scientists. All key science over the last hundred years has been reported in journals, and subjected to peer review. Unlike religion, contemporary science is almost entirely independent of the particular culture in which it is done: science is universal, and there is no Western or Eastern science. The belief that scientific ideas are continually changing is true mainly at the frontiers of investigation, but the core is largely solid. It is fundamental to science that even the most deeply held beliefs about science, from Einstein to Darwin, can in principle be shown to be faulty and so require modification. It is also important to realise that reliable scientific beliefs have no intrinsic ethical or moral content: they refer to how the world is. There are no ethics in Newton's laws, nor in the genetic code, nor in the fact that genes can affect our mental health.

I want to emphasise one common feature of science that I think relevant to understanding why non-scientists can have severe difficulty with scientific beliefs, that is, the unnatural nature of science. Science does involve a special way of thinking about the nature of the world. Instead of just looking at the

obvious relations between cause and effect, like a stone breaking due to the force of a hammer, science tries to understand
mechanisms at a deeper level, and this leads to serious problems,
for the way the world works does not fit with our day-to-day
expectations, our common sense. Almost without exception,
any common-sense view of the world is scientifically false.
Obvious examples are the movement of the sun with respect to
the earth, and the fact that a force on a body does not cause
movement, but acceleration. How unnatural it is to believe,
rightly, that at a constant 400 miles per hour on a plane, there
is no force acting on you to move you ahead. And how well
does Darwin's theory of evolution by random variation and
natural selection fit with common sense? Even the number of
molecules in a glass of water is beyond anyone's natural expectation: there are more than there are glasses of water in the
oceans. No matter where one looks in science, its ideas con-
found common sense. It is not even easy to think of how ice
cools one's drink in the correct way: cold does not flow from
the ice to the liquid, it is the heat flowing from the liquid that
melts the ice.

And things get much worse when one enters the world of
subatomic particles, quantum mechanics, black holes and the
Big Bang. Everyday analogies completely break down. Science
can be very difficult. I believe in the Big Bang, while I do not
understand the evidence, but I think I could understand it if I
took off, say, five years to study the physics. Part of the problem
is that much of science is based on mathematics, which can be
very alienating because it is so technical, and often counterintuitive. There are also complex instruments that are used to
make all sorts of measurements.

Another feature of science, which is rarely used in day-to-
day life, is statistical and mathematical analysis of the likelihood of events; for example, common-sense estimates of

probability can be wildly inaccurate, as we have seen. Even in biology, it is not easy to understand how genes control the behaviour of the cells by coding for the proteins that are the key agents in cell behaviour. Surveys in Europe have even found that about half the population believed that tomatoes did not have genes unless they had been genetically modified.

Common sense thus does not lead to science. Doing science requires a special self-awareness, and it is often necessary to resist common sense, since an unfamiliar quantitative rigour is required. Indeed, one can live one's life rather well without knowing any science at all, since most of it has little direct relevance to day-to-day events. Sherlock Holmes's response to Watson's criticism of his ignorance of science was: 'What the devil is it to me if you say we go round the sun? If we went round the moon it would not make a pennyworth of difference to me.' Of course, people like Holmes are excluding themselves from the greatest intellectual achievement of our age.

One should treat with caution anything the philosophers of science have to say about the nature of scientific belief, and that includes Karl Popper with his emphasis on falsification as the key feature of science. There is one particular set of beliefs about science that needs careful examination and ultimate rejection: that is the relativist view of science. These beliefs are held by a number of influential sociologists of science who have promoted a portentously named programme: the Strong Programme of the Sociology of Science. They claim that science is no more than a social construct, another special set of myths and beliefs that give the illusion of truth, and so are little different at their core from other beliefs about how the world works. Rejection of this claim does not mean that science is not a social process: scientists are all too aware of the problems of getting grants, getting papers accepted by leading journals, and cooperation and also competition with other scientists. I do not understand how these sociologists of science can hold such an absurd set of beliefs, and can only suggest envy, for sociology cannot compare its progress with that in the 'harder' sciences, from physics to even psychology. Postmodernists promote relativism and structuralism so that reality almost ceases to exist – everything is a social construction, including science. Their beliefs so influenced their thinking that the hoax article using their kind of language was regarded as genuine and important.

Belief in a scientific idea by scientists is not free of emotion or prejudice. We love our own ideas, and treat evidence that contradicts them, at least initially, with great suspicion if not hostility. As a scientist, one should not give up one's beliefs too easily, for, as has been suggested, the graveyard of failed scientists is filled with those who gave up their ideas at the first bit of evidence that they were wrong – that evidence can itself be faulty. Also, as has been remarked by another great scientist, the physicist Max Planck, 'A new scientific truth does not triumph by convincing its opponents and making them see the light, but rather because its opponents generally die, and a new generation grows up that is familiar with it.' A bit strong, but basically correct.

Scientists can be faulty with their own beliefs, as well as those of other scientists, until the evidence is persuasive. Consider Alfred Wegener in the 1920s, who put forward the idea that the continents of Africa and South America had once been joined together, but over millions of years had drifted apart. There was tremendous hostility to these ideas and it was only in the 1960s that new evidence, based on measurements of the earth's magnetic field, gave his ideas strong support. Rather different is the case of the great physicist Lord Kelvin at the end of the nineteenth century. He would not accept the suggestion that the age of the earth could be of the order of millions of years. His opposition was based on the data relating to the cooling of the earth, but it was only later that it became clear that radioactivity heated the earth, and so made his calculations completely wrong.

Isaac Newton was also faced with criticism similar to that currently made about the paranormal. When he put forward his theory of gravity, that all bodies attracted each other with a force proportional to their mass and inversely proportional to the distance between them, it was according to the great Leibniz a return to occult qualities; and others said that it was totally inexplicable by the current knowledge of mechanics. Newton replied that yes, it seemed to him a great absurdity that one body could act on another at a distance through a vacuum without the mediation of something else. But, and this is crucial, the evidence was overwhelming, and he admitted that he had been unable to deduce the reason for these properties, but said 'I do not feign hypotheses.'

There is also, on occasion, what has been called pathological science, a term coined by the chemist Irving Langmuir. Langmuir was referring to the scientific claims for phenomena like telepathy and other examples of extrasensory perception, which few scientists now take seriously. His criteria for pathological science are that the maximum effect is very small, often near the limit of detectability; the magnitude is independent of the cause; there are claims of extreme accuracy; there is a fantastic theory; and criticisms are met with ad hoc excuses. A recent example is possibly that of the claim for cold fusion providing large quantities of energy at very low cost. The claims have now been dismissed, and one does not know to what extent the researchers just made mistakes, or whether fraud was involved in this and other cases of pathological science.

Science is not the same as technology. Technology alters nature: things are made. The final product of science is understanding, while that of technology is a product, something that is used. Much of modern technology is based on science, but this link is of recent origin, since science had virtually no impact on technology until the nineteenth century. Technology includes the ancient arts of agriculture and metal making, as well as the great Renaissance buildings and the machines and engines of the Industrial Revolution. All these were achieved without any influence from science, but were based on causal beliefs. The steam engine owed almost nothing to science – it probably could have been built by the Greeks. It is possible to have complex technology without any scientific understanding at all, but there must be a basic understanding of cause and effect. It was only in the late eighteenth century that science influenced technology, and its impact has increased enormously since then. The relationship between science and technology is not symmetrical, since technology itself had an enormous impact on science, which could not have advanced without it.

All science, as we know it, comes from the Greeks. Thales of Miletus seems to have been one of the very first scientists as he tried to explain the world, not in terms of myths, but in more concrete terms, terms that might be subject to verification. What, he wondered, might the world be made of? His unexpected answer was: water. Water could clearly change its form from solid to liquid to gas and back again; clouds and rivers were in essence water; and water was essential for life. His belief was fantastical perhaps – contrary to common sense – but so is the essence of science. But more important than his answer was his explicit attempt to find a fundamental unity in nature. It expressed the belief that underlying all the varied forms and substances in the world, unifying principles and causes could be found. The possibility of objective and critical thinking about nature had begun. Never before, it seems, had someone put forward non-mystical ideas about the nature of the world that might be universal explanations, and there was for the first time a conviction that the natural world was controlled by what might loosely be called 'laws'.

Thales' idea was open for discussion and debate. It was a wonderful leap to free thinking from the straitjacket of mythology, and the grip of relating everything to man. Here, for the first time, attention, and the resulting beliefs, were focused on the nature of the world with no immediate relevance to humankind. Human curiosity had hitherto been entirely devoted to man's relation to nature, and not to nature itself. With the Greeks, man and nature were for the first time no longer believed to be inextricably linked, and there begins to be a distanced curiosity about the world itself.

Perhaps this change is a reflection of the nature of Greek society, with its own traditions in law and literature. Their citizens cared about evidence. Since science had very few practical applications, it was the love of knowledge that drove the Greeks to the philosophy that led to science. And this, as Aristotle made clear, required leisure, in which slaves may have played a key role.

It was also Thales who established mathematics as a science, irrespective of how much he might have learned from the Babylonians and Egyptians, who had established arithmetic procedures and the elements of geometry for their practical needs. The Babylonians knew elements of geometry as early as 1700 BC and had tables listing the sides of right-angled triangles, so they must have been aware of the key features of Pythagoras's theorem, which states that the square of the hypotenuse is the sum of the squares of the other two sides. Thales turned these tools of measurement into a science. He put forward a number of basic propositions: that a circle is bisected by its diameter; that, if two straight lines cut each other, the opposite angles are equal; and that the angle inscribed in a semicircle is a right angle. Here, for the first time, were general statements about lines and circles – statements of a kind never made before. They were true beliefs that applied to all circles and lines everywhere, and that is the generality to which science aspires. The Greeks, with Euclid as the exemplar, transformed a varied collection of empirical rules for calculation into an ordered abstract system. Mathematics was no longer merely a tool used for practical problems; it became a science.

Aristotle believed in four types of causes: material, that of which a thing is made; formal, its characteristic features; final, its functions; and efficient, how it came about. Only the efficient cause fits with our idea of cause. Even the Greeks had debates as to whether final causes provided any valid explanation of events and things: teleology. When Aristotle first formulated the principles of non-contradiction, he may have aimed to provide explicit rules for what he may have considered was already implicit in human communication. Euclid's *Elements* later provided a majestic model for logical deduction. But before the Greeks, the concept of contradiction was just not there, and so it was possible to hold totally contradictory beliefs.

The concept of proof is central to some aspects of belief. This is an idea that had its origins in Greece in the sense that they made the concept explicit. Aristotle showed, and he was the first, that a proof depended both on deductive argument, and the validity of the premises from which the deductions were made. If the premises were true, one could believe in the deductions made from them. The practice of proof was used among other ancient societies, but the explicit concept is Greek alone.

By contrast, the Chinese laid the emphasis on dialectic rather than formal logic. They did not engage in speculative reasoning for its own sake. The Chinese had wonderful technology, but no science, and their beliefs were quite mystical. There was nothing like the Greek discussion of cause. The three realms of heaven, earth and man were united in the person of the emperor who mediated between them. The human body was believed to be a microcosm of the whole world, heaven and earth. And as regards law, litigation was virtually unknown, and issues were settled by officials or by discussion. The Chinese were, in general, very much less confrontational with respect to ideas, and even recognised that silence would be superior to words. It was also the irregular that interested them; there was little interest in explaining the regular. The Mohists argued that what eyewitnesses reported they had seen or heard should, in general, be accepted; and this was the basis for their acceptance of the existence of ghosts and spirits. Mo Zi, who lived in China in the fourth century BC and founded Mohismput it like this:

> The way to find out whether anything exists or not is to depend upon the testimony of the eyes and ears of the multitude. If some have heard it or some have seen it then we have to say it exists. If no one has heard it and no one has seen it then we have to say it does not exist. So why not go to some villages or districts and inquire? If from antiquity to the present, and since the beginning of man, there are men who have seen the bodies of ghosts and spirits, and have heard their voices, how can we say that they do not exist?

Most of Greek science turned out to be wrong – from Aristotle's ideas about motion to his embryology. His most important contribution was that of setting up postulates and drawing logical conclusions from them, and this was brilliantly exploited by both Euclid and Archimedes. Archimedes is probably the first scientist to apply mathematics to physical phenomena correctly, and so is probably the first applied mathematician, and his contributions to the mechanics of levers and hydrostatics remain true today. This is a nice counter-example to those who believe that scientific ideas are constantly being shown to be wrong and replaced.

When applying mathematics to nature, Archimedes realised

that the system had to be made a bit unreal. For example, in considering the balancing of a lever, the composition of the lever and friction were irrelevant. Archimedes' analysis of the simple balance and the relationship between weights and their distance from the fulcrum is remarkably ingenious. His analysis of floating bodies is also startlingly original, but can be thought of as a continuation or extension of the work on levers, since it is essentially about the analysis of forces at equilibrium. Just consider the achievement of the belief stated in his second postulate: 'Let it be granted that bodies which are forced upwards in a fluid are forced upwards along the perpendicular to the surface which passes through their centre of gravity.' From such postulates, he showed that the loss of weight of a body in a fluid is equal to the weight of water displaced, and he went on to discover the specific gravity of substances. To see just how unnatural even simple science can be, consider the following problem, bearing in mind that Archimedes would have laughed at its simplicity. Imagine that you are in a rubber dinghy floating in a swimming pool. There is a large bag of stones with you in the boat. You throw the bag into the water and it sinks. Do you believe that the level of the water in the pool goes up, goes down, or remains the same? It goes down.

The limited use of experiments of any sort, and the failure to develop the experimental method, are curious features of Greek science. These were advances that evidently required a special mode of thought, which only came about in the sixteenth and seventeenth centuries. The fall of a body is a good example. No one really challenged Aristotle's view that the rate of fall of a body is proportional to its weight, even though Philoponus (sixth century AD) did do some experiments that proved it to be wrong. It was not until Galileo's time that the intellectual environment permitted a demonstration that Aristotle's view was both logically inconsistent and experimentally false. Galileo transformed physics and gave much credit to Archimedes.

Science is basically in conflict with religion, for there is no scientific evidence for any gods or their supposed powers and special forces. Yet many scientists have been and are religious. Isaac Newton is an outstanding example of a very great scientist who was also deeply religious. 'We must believe there is one God or supreme Monarch . . .' While denying that Christ was consubstantial and co-eternal with the Father, Newton asserted that Christ deserved worship as the Redeemer. There is evidence from his manuscripts that he grounded his law of the universality of gravitation on the omnipresence of the one God whose free will determined 'the frame of the world'. For him, it seems, the very beauty of the solar system reflects the aesthetic sensibility of God the Creator. The law of gravity did not have to be an inverse square law had God not wished it to be. He also saw in alchemy a legitimate instrument for the imitation of the deity. And in India today there are strong claims that Vedic Hinduism is simply science by another name.

Darwin specifically said that he avoided writing about religion and confined himself to science. His ideas, however, are probably the most important in all of science to challenge some religious beliefs. The idea that humans have evolved from very simple and primitive creatures by random changes and natural selection is intolerable to those who believe, for example, that humans were created by God as described in the Bible. Some 65 per cent of Americans believe that human beings were directly created by God, while only 22 per cent believe in Darwinian evolution and intelligent design. This indicates a big swing towards creationist beliefs over the last ten years. Curiously, it is partly with so-called scientific arguments that ardent creationists attack Darwinian evolution.

This is not to say that all the scientific questions relating to evolution have been solved. On the contrary, the origin of life itself, the evolution of the miraculous cell from which all living beings evolved, is still poorly understood. In his *Origin of Species*, Darwin was clearly aware of the unnatural nature of science, for he wrote:

> Nothing at first can appear more difficult to believe than that the more complex organs and instincts should have been perfected not by a means superior to, though analogous with, human reason, but by the accumulation of innumerable slight variations, each good for the individual possessor.

There is no plan, no intelligence. And George Bernard Shaw's heart sank when he realised this. 'There is a hideous fatalism about it.'

Sam Berry, a zoologist, points out that in relation to miracles, credibility should be kept separate from mechanism. Very rarely does the Bible confide the methods God used to achieve his purpose. One exception is in Exodus, where 'the Lord drove the sea back by a strong east wind all night, and made the sea dry land, and the waters were divided. And the people of Israel went into the midst of the sea on dry ground . . .' The actual site is not known, but there are sites where it may have been possible for the sea to part under certain weather conditions. Again, the plagues of Egypt could have been due to natural illness caused by flies and other insects, and not just used by God as a miracle to help his people.

The virgin birth of Jesus is, for Berry, however, in a different class: the son of God born to a virgin, Mary. This was part of Christian belief from the early centuries, but some Christians have had difficulty with it, not least since the alternative would be an illegitimate birth. In 1985 the bishops of the Church of England concluded that the evidence from Matthew and Luke was not detailed or decisive and the decision to believe in the virgin birth of Jesus '. . . has to be a matter of faith'. The doctrine of the Bible, ought, Berry claims, to be sufficient. It is necessary, however, to recognise that all the research on the embryos of humans and related animals has found that an egg cannot develop normally without a sperm. Moreover, a woman cannot without a male sperm bear a male child. Yet, Berry argues, it is not impossible for the virgin birth to have taken place, given a whole set of improbable conditions in Mary and the embryo.

But what do other scientists believe about religion? In 1914 and 1933 there were surveys of scientists in which they were asked whether they believed in a God who has communication with humans, and to whom one might pray and expect an answer, and whether they believed in personal immortality. Yes, no, and don't know were the only permitted answers. Similar surveys were repeated in 1996 and 1998. Over all those years, from 1914 to 1998, the proportion of scientists believing in God remained at around 40 per cent. However, for scientists of distinction – the scientific elite – only 30 per cent believed in God in 1914, and by 1933 it was only 20 per cent. In 1998 the elite scientists from the prestigious National Academy of Science in the USA were overwhelmingly 90 per cent non-believers, with biologists being most strongly represented. While some 45 per cent of ordinary Americans reject Darwinian evolution, in the UK the figure is around 10 per cent.

There are some curious features relating science to religion. Higher education in the USA has some 1,000 courses on science and faith, and the Templeton Foundation supports courses and meetings on this topic. It is an attempt to reconcile religious belief with modern science. The National Academy of Science issued a report promoting the teaching of evolution in schools, which begins: 'Whether God exists or not is a question about which science is neutral.' There is also something of a thaw in the icy boundary between science and religion. The Vatican has apologised for its treatment of Galileo, and Pope John Paul II acknowledged that evolution may be more than just a hypothesis. The Nobel Laureate in physics, Charles Townes, has written: 'The more we know about the cosmos and evolutionary biology, the more they seem inexplicable without some aspect of (intelligent) design and for me that inspires faith.' By contrast, Richard Dawkins, the evolutionary biologist, is an opponent of such spiritual thought:

> The universe we observe has precisely the properties we would expect if there is at bottom no design, no purpose, no evil, no good, nothing but pointless indifference . . . We are machines for propagating DNA . . . It is every living object's sole reason for living.

Anyone, he says, who still believes in a Creator God is scientifically illiterate. But Francis Collins, director of the Human Genome Research Institute at the National Institutes of Health, reported the following experience when a new gene was revealed:

> . . . a feeling of awe at the realisation that humanity now knows something only God knew before . . . A lot of scientists really don't know what they are missing by not exploring their spiritual feelings.

And why, he asks, should God not have used the mechanisms of evolution?

William James was unpersuaded that science could provide all the answers, and in one sense he is right, for however much we understand, there will always be unanswered questions. So even if string theory provides an explanation for the laws of physics, and everything is essentially reducible to these laws, we are still left with the problem of where the strings came from, as well as their properties. We must have the intellectual courage to live with such unanswered questions, rather than invent answers that have no basis other than in mystical experience. But we must also accept that science can tell us nothing about ethics or morality. As James says, the saintliness of religion gives the world indispensable qualities and is one way of judging how we should behave. But it is certainly not the only way, as we humanists know.

Of course, it is possible for God to easily reveal to scientists his current existence: God only has to perform, publicly, one or two miracles, for good evidence to be provided. This evidence could, for example, be quite simple, like turning a lake into good red wine, or providing an instant cure for cancer. Such miracles would almost certainly lead to religious beliefs among the sceptics.

Science provides by far the most reliable method for determining whether one's beliefs are valid. It may be difficult, as it will go often against common sense, but its value is inestimable.

CHAPTER THIRTEEN

Believable?

> It is undesirable to believe a proposition when there is
> no ground whatever for supposing it is true.
>
> Bertrand Russell

It may be hard to believe that religion and causal beliefs in general had their evolutionary origin in tool making, which drove human evolution. But that is what I have attempted to argue in the preceding chapters. The evidence is fragmentary, but I know of no good evidence that contradicts this evolutionary story. I am fully aware of how complex the brain is, and it has not been possible for me to account for the changes in evolution in terms of neural processes. I have purposely avoided any discussion of consciousness, which still remains mostly poorly understood. When we really understand how the brain functions, particularly the circuits involved in creating beliefs and so controlling movements, all will be so much clearer and more interesting.

Linking religion to tool use may, I regret, offend those with religious beliefs. The suggestion that religious beliefs may be genetically programmed in our brains will also offend those who are religious, and may even irritate those who are not. By genetically programmed, I mean that there are circuits in our brain that are set up by the genes that predispose us to have religious and mystical beliefs. It is hard to imagine that the religious and mystical beliefs found in every culture have some other origin. The detailed nature of these beliefs is influenced by the local history and culture. Moreover a case can be made that some of the false beliefs that arise in mental illness, particularly those with a mystical quality, can be linked to the religious circuits in the brain. How else could one account for how common mystical experiences are in the population, or the effects of LSD?

Less controversial, perhaps, is the idea that tool use is closely linked to the evolution of causal beliefs. No doubt language evolution played an important role, as did culture. It remains a puzzle why other primates did not evolve to make complex tools as they are on the edge of having causal beliefs about the physical world. Understanding the physical interaction of objects is quite unlike understanding social behaviour, how others like oneself will behave. But why did it evolve in humans? Perhaps they needed something to give them an advantage over stronger and more agile primates.

Whatever its evolutionary history, we all have a belief engine that works continuously to provide us with causal explanations of events that affect our lives. Availability and representativeness, as the basis on which many beliefs are constructed, could have been useful in evolution to help us avoid danger. Many causal beliefs serve our daily lives very well, but there are many that are very unreliable, particularly, for example, in relation to risk. One also needs to note how when there is an accident like a plane crash, the relatives of those who died are obsessed with trying to find the cause. There are also those bizarre paranormal beliefs, essentially mystical, about, for example, special life energy in healing, and telepathy. On the whole, such beliefs are rarely life-threatening, but they can undermine confidence in science-based medical treatments, with serious results.

An important issue is the relationship between scientific and religious beliefs. I would argue that people have the right to hold whatever beliefs appeal to them, but with a fundamental provision that those beliefs must be reliable if they lead to actions that affect the lives of other people. I have never tried to persuade my religious son not to be religious, as he has benefited greatly from his religious beliefs, and the religious group to which he belonged. While religious and scientific beliefs are fundamentally different, it is only when they come into conflict that real problems arise. Stem cells and the status of the human embryo is a current example.

There are innumerable areas where the beliefs of others affect our lives, and this is where the dangers lie. Commerce and industry and technology are based on complex beliefs. In Britain, those against GM crops increased at one stage to nearly 90 per cent largely because it was seen as an unacceptable interference with nature and because it was profit driven. Also there seemed to be no benefits for the public. It is a case where people look for benefits, and believe those whom they trust, and government and big business do not command high levels of trust. How should we respond to claims about possible global warming? There are well-argued cases that say implementing the Kyoto agreement will be very costly and will only result in a small delay in the warming. How to treat the environment is an area surrounded by different beliefs, over which individuals have little control. Other examples include the low status of women in certain societies that are determined by religious beliefs. Beliefs in the moral sphere, particularly in relation to politics and religion, present a major problem, as they can lead to disaster and countless deaths.

What of the future? The beliefs of science are the most reliable we have about how the world works, so could these, with time, become the everyday basis for forming beliefs? I very much doubt it. Science, as I have argued, goes against common sense, and we also usually lack the necessary information on which to make a scientific judgement. But more importantly, our belief engine, programmed in our brains by our genes, operates on different principles. It prefers quick decisions, it is bad with numbers, loves representativeness, and sees patterns where often there is only randomness. It is too often influenced by authority, and it has a liking for mysticism.

Religious and mystical beliefs will continue for the foreseeable future to be held by millions of people, not only because mysticism is in our brains, but also because it gives enormous comfort and meaning to life. And it provides a basis for causal beliefs about fundamental human issues. Just look at the strength of religion in an advanced industrial culture like the USA. And while we may be hostile to the beliefs of others, we need always to remember that it is having beliefs that makes us human. We have to both respect, if we can, the beliefs of others, and accept the responsibility to try and change them if the evidence for them is weak or scientifically improbable. The loss of religious beliefs could have very serious consequences, and so could the enforcement of those beliefs on others. It is the action based on beliefs that ultimately matters, and respect for the rights of others is fundamental.

I can only conclude by quoting Virgil, from the first century BC: 'Happy he, who can understand the causes of things.'

References

ONE **Everyday**

Adams, J. (1995) *Risk*. Routledge, London.

Furnham, A. F. (1988) *Lay Theories*. Pergamon, Oxford.

Furnham, A. and Schofield, S. (1987) Accepting personality test feedback: a review of the Barnum Effect. *Curr. Psych. Res. Review*, 6,
162–78.

Geertz, C. (1993) *Local Knowledge*. Fontana, London.

Goel, V. and Dolan, R. J. (2003) Explaining modulation of reasoning
by belief. *Cognition*, 87, B11–B22.

Myers, D. G. (2002) *Intuition: Its Powers and Perils*. Yale University
Press, New Haven.

Nisbett, R. E. et al. (2001) Culture and systems of thought: holistic
versus analytic cognition. *Psychological Rev.*, 108, 291–310.

Roud, S. (2003) *The Penguin Guide to Superstitions of Britain &
Ireland*. Penguin, London.

Schumaker, J. F. (1995) *The Corruption of Reality*. Prometheus, New
York.

Shermer, M. (1999) *How We Believe*. Freeman, New York.

Sutherland, S. (1992) *Irrationality: The Enemy Within*. Constable,
London.

TWO **Belief**

Barlow, J., Cosmides, L. and Toobey, J. (1992) *The Adapted Mind:
Evolutionary Psychology and the Generation of Culture*. Oxford
University Press, Oxford.

Blackmore, S. (1999) *The Meme Machine*. Oxford University Press, Oxford.

Churchland, P. S. (2002) Self-representation in nervous systems. *Science*, 296, 308–10.

Hume, D. (1975) *Enquiries Concerning Human Understanding and Concerning the Principles of Morals*, Oxford University Press

Needham, R. (1972) *Belief, Language, and Experience*. Blackwell, Oxford.

Pinker, S. (1997) *How the Mind Works*. Norton, New York.

Schacter, D. L. and Scarry, E. (eds) (2000) *Memory, Brain, and Belief*. Harvard University Press, Cambridge MA.

THREE Children

Bibace, R. and Walsh, M. E. (1980) Development of children's concept of illness. *Pediatrics*, 66, 912–17.

Callanan, M. A. and Oakes, L. M. (1992) Preschoolers' questions and parents' explanations: causal thinking in everyday activity. *Cognitive Development*, 7, 213–33.

Corrigan, R. and Denton, P. (1996) Causal understanding as a developmental primitive. *Dev. Rev.*, 16, 162–202.

Dunn, J. (2000) Emotion and the development of children's understanding. *European Review*, 8, 9–15.

Gopnik, A. et al. (1999) *How Babies Think*. Weidenfeld, London.

Harris, P. L. (2000) *The Work of the Imagination*. Blackwell, Oxford.

Karmiloff, K. and Karmiloff-Smith, A. (1999) *Everything Your Baby Would Ask*. Golden Books, New York.

Karmiloff-Smith, A. (1992) *Beyond Modularity*. MIT Press, Cambridge, MA.

Leslie, A. M. (1984) Infant perception of a manual pick-up event. *Brit. J. Devel. Psychology*, (19–32).

Piaget, J. (2001) *The Child's Conception of Physical Causality*. Transaction, New Jersey.

Schlottmann, A. (2000). Is perception of causality modular? *Trends in.Cognitive Sciences*, 4, 441–2.

Schlottmann, A. (2001). Perception versus knowledge of cause-and-effect in children: when seeing is believing. *Current Directions in Psychological Science*, 10, 111–15.

Schlottmann, A. et al. (2002). Children's intuitions of perceptual causality. *Child Development*, 73, 1656–77.

Sperber, D. et al. (eds) (1995) *Causal Cognition: A Multidisciplinary Debate*. Clarendon Press, Oxford.

Van der Meer, A. L. et al. (1995) The functional significance of arm movements in neonates. *Science*, 27, 293–5.

FOUR **Animals**

Cacchione T, and Krist, H. (2004) Recognizing impossible object relations: intuitions about support in chimpanzees (Pan troglodytes). *J. Comp. Psychol.*, 118, 140–8.

Chappell, J. and Kacelnik, A. (2002) Tool selectivity in a non-primate, the New Caledonian crow (Corvus moneduloides). *Animal Cognition*, 5, 71–8.

Dunbar, R. (2004) *The Human Story*. Faber and Faber, London.

Emery, N. J. and Clayton, N. S. (2004) The mentality of crows: convergent evolution of intelligence in corvids and apes. *Science*, 306, 1903–7.

Hauser, M. (2000) *Wild Minds*. Allen Lane, London.

Hauser, M., Pearson, H. and Seelig, D. (2002) Ontogeny of tool use in cottontop tamarins, Saguinus oedipus: innate recognition of functionally relevant features. *Animal Behaviour*, 64, 299–311.

Hauser, M. D. (2001) Elementary, my dear chimpanzee. *Science*, 291, 440–1.

Hayes, L. and Huber, L. (eds) (2000) *The Evolution of Cognition*. (Many key articles including Tomasello, Dunbar.) MIT Press, Cambridge, MA.

Povinelli, D. J. (2000) *Folk Physics for Apes*. Oxford University Press, Oxford.

Premack, D. and Premack, A. (2002) *Original Intelligence*. McGraw-Hill, New York.

Tomasello, M. (1999) *The Cultural Origins of Human Cognition*. Harvard University Press, Cambridge, MA.

Tomasello, M. et al. (2003) Chimpanzees understand psychological states –the question is which ones and to what extent? *Trends in Cognitive Science*, 7, 153.

Whiten, A. (2000) Primate culture and social learning. *Cognitive Science*, 24, 477–508.

FIVE **Tools**

Ambrose, S. H. (2001) Paleolithic technology and human evolution. *Science*, 291, 1748–52.

Calvin, W. H. (1993) The unitary hypothesis: A common neural circuitry for novel manipulations, language, plan-ahead and throwing. In: K. R. Gibson and T. Ingold (eds) *Tools, Language, and Cognition in Human Evolution*, pp. 230–50. Cambridge University Press, Cambridge.

Calvin, W. H. (2003) *A Brief History of the Mind: From Apes to Intellect and Beyond*. Oxford University Press, Oxford.

Calvin, W. H. and Bickerton, D. (2000) *Lingua ex Machina: Reconciling Darwin and Chomsky with the Human Brain*. MIT Press, Cambridge, MA.

Corballis, M. C. (2002) *From Hand to Mouth: The Origins of Language*. Princeton University Press, Princeton, NJ.

Corballis, Michael C. and Lea, Stephen E. G. (eds) (1999) *The Descent of Mind: Psychological Perspectives on Hominid Evolution*. Oxford University Press, Oxford.

Donald, M. (1991) *The Origins of the Modern Mind*. Harvard University Press, Cambridge, MA.

Dunbar, R. (1996) *Grooming, Gossip and the Evolution of Language*. Faber and Faber, London.

Gibson, K. R. and Ingold, T. (1993) *Tools, Language and Cognition in Human Evolution*. Cambridge University Press, Cambridge.

Greenfield, P. M. (1991) Language, tools and the brain. *Behavioural & Brain Sciences*, 14, 531–95.

Holden, C. (2004) The origin of speech. *Science*, 303, 1316–19.

Lieberman, P. (2000) *Human Language and our Reptilian Brain*. Harvard University Press, Cambridge, MA.

Johnson-Frey, S. H. (2003) The neural basis of complex tool use in humans. *Trends. Cog. Sci.*, 8, 71–8.

Johnson-Frey, S. H. (2003) What's so special about human tool use? *Neuron*, 39, 201–4.

Mithen, S. (1996) *The Prehistoric Mind: The Cognitive Origins of Art and Science*. Thames & Hudson, London.

Oakley, K. (1949) *Man the Toolmaker*. British Museum Press, London.

Ofek, H. (2002) *Second Nature: Economic Origins of Human Evolution*. Cambridge University Press, Cambridge.

Premack, D. (2004) Is language the key to human intelligence? *Science*, 303, 318–20.

Schick, K. D. and Toth, N. (1993) *Making Silent Stones Speak*. Weidenfeld & Nicolson, London.

Tomasello, M., Carpenter, M., Call, J., Behe, T. and Moll, H. (2004) Understanding and sharing intentions: the origins of cultural cognition. *Behavioural and Brain Sciences*, 28, 721–27

Vogel, G. (2002) Can chimps ape ancient hominid toolmakers? *Science*, 296, 1380.

Washburn, S. (1978) The evolution of man. *Scientific American*, 219, 146–55.

Wilson, F. O. (1998) *Consilience*. Little Brown, London.

Wilson, F. R. (1999) *The Hand*. Vintage, New York.

Wolpert, L. (2003) Causal belief and the origins of technology. *Philos. Transact. Royal Soc.*, A 361, 1709–19.

Wynn, T. (1996) The evolution of tools and symbolic behaviour. In: A. J. Lock and C. R. Peters (eds), *Handbook of Human Symbolic Evolution*, 263–87. Clarendon Press, Oxford.

SIX Believing

Geertz, C. (1983) *Local Knowledge*. Basic Books, New York.

Gilovich, T. (1991) *How We Know What Isn't So*. Free Press, New York.

Gilovich, T., Griffin, D. and Kahneman, D. (2002) *Heuristics and Biases: The Psychology of Intuitive Judgement*. Cambridge University Press. Cambridge.

Myers, D. G. (2002) *Intuition*. Yale University Press, New Haven, CT.

Roth, H. (2001) *Change, Choice and Inference: A Study of Belief, Revision and Nonmonotonic Reasoning*. Clarendon Press, Oxford.

Runciman, G. (1991) Are there any irrational beliefs? *Archives European Sociology*, 23, 215–28.

Schumaker, J. F. (1995) *The Corruption of Reality*. Prometheus, Amherst, NY.

SEVEN **False**

Black, S. et al. (1963) Inhibition of Mantoux reaction by direct suggestion under hypnosis. *British Medical Journal*, 1, 1649–52.

Blakemore, S. J., Oakley, D. A. and Frith, C. D. (2003) Delusions of alien control in the normal brain. *Neuropsychology*, 41, 1058–67.

Burgess, P. W. and Shallice, T. (1996) Confabulation and the control of recollection. *Memory*, 4, 359–411.

Colbert, S. M. and Peters, E. R. (2002) Need for closure and jumping-to-conclusions in delusion-prone individuals. *Journal of Nervous and Mental Disease*, 190, 27–31.

Coltheart, M. and Davies, M. (eds) (2002) *Pathologies of Belief*. Blackwell, Oxford.

Frith, C. D., Blakemore, S. and Wolpert, D. M. (2000) Explaining the symptoms of schizophrenia: abnormalities in the awareness of action. *Brain Research Review*, 31, 357–63.

Gelder, M. et al. (2001) *Shorter Oxford Textbook of Psychiatry*. Oxford University Press, Oxford.

Goodman, N. (2002) The serotonergic system and mysticism: could LSD and the nondrug-induced mystical experience share common neural mechanisms? *Journal of Psychoactive Drugs*, 34, 263–72.

Johnson, M. K. and Raye, C. L. (1998) False memories and confabulation. *Trends in Cognitive Sciences*, 2, 136 –45 .

Kosslyn, S. M. et al. (2000) Hypnotic visual illusion alters color processing in the brain. *American Journal of Psychiatry*, 157, 1279–84 .

Lee, M. A. and Shlain, B. (1985) *Acid Dreams*. Grove, New York .

McKay, R., Langdon, R. and Coltheart, M. (2005) Sleights of mind; delusions, defences and self-deception. *Cognitive Neuropsychiatry*, 10, 305–26

Nash, M. R. (2001) The truth about hypnosis. *Sci. American,* July, 37–43.

Noble, J. and McConkey, J. M. (1995) Hypnotic sex change: creating and challenging a delusion in the laboratory. *Jnl. Abnormal. Psychol.*, 104, 69–74.

Persaud, R. (2003) *From the Edge of the Couch*. Bantam, London.

Peters, E. et al. (1999) Delusional ideation in religious and psychotic populations. *British Journal of Clinical Psychology*, 38, 83–96.

Ramachandran, V. S. and Blakeslee, S. (1998) *Phantoms in the Brain*. Fourth Estate, London.

Schumaker, J. F. (1990) *Wings of Illusion*. Prometheus, Amherst, NY.

Shermer, M. (2000) *How We Believe*. Freeman, New York.

Wells, B. (1973) *Psychedelic Drugs*. Penguin, London.

EIGHT Religion

Alper, M. (2000) *The 'God' Part of the Brain*. Rogue Press, New York.

Armstrong, K. (1999) *A History of God*. Vintage, London.

Barker, E. (1999) *New Religious Movements: A Practical Introduction* (2nd edition). HMSO, London .

Burket, W. (1996) *Creation of the Sacred: Tracks of Biology in Early Religions*. Harvard University Press, Cambridge, MA.

Easterbrook. G. (1997) Science and God: A warming trend? *Science*, 277, 890–3.

Habgood, J. (2000) *Varieties of Unbelief*. Darton, Longman and Todd, London.

Hinde, R. (1999) *Why Gods Persist*. Routledge, London.

James, W. (1897, 1956) *The Will to Believe*. Dover, New York.

James, W. (1902, 1982) *The Varieties of Religious Experience*. Penguin, London.

Kendler, K. S. et al. (2003) Dimensions of religiosity and their relationship to lifetime psychiatric and substance abuse disorders. *American Journal of Psychiatry*, 160, 496–502.

Koenig, H. G. et al. (2001) *Handbook of Religion and Health*. Oxford University Press, Oxford.

Larson, E. J. and Witham, L. (1999) Scientists and religion in America. *Scientific American*, September, 78–82.

McKay, R. (2004) Hallucinating God? *Evolution and Cognition*, 10, 1–12.

Newberg, A., D'Aquila, E. and Radic, V. (2001) *Why God Won't Go Away*. Ballantine, New York.

Smart, N. (1998) *The World's Religions* (2nd edition). Cambridge University Press, Cambridge.

Stark, R. and Finke, R. (2000) *Acts of Faith*. University of California Press, Berkeley.

Wilson, D. S (2002) *Darwin's Cathedral*. University of Chicago Press, Chicago.

Wright, L. (2002) Lives of the Saints. *New Yorker*, 21 January, 4.

NINE **Paranormal beliefs**

Alcock, James E. (1995) The belief engine. *Skeptical Inquirer*, May/June, 14–18.

Brigg, R. (1996) *Witches and Neighbours*. Viking, London.

Brugger, P. and Taylor, K. I. (2003) ESP: Extrasensory perception or effect of subjective probability? *Journal of Consciousness Studies*, 10, 221–28

Brugger, P. (2001) From haunted brain to haunted science: a cognitive neuroscience view of paranormal and pseudoscientific thought. In: J. Houran and R. Lange (eds), *Hauntings and Poltergeists: Multi-disciplinary Perspectives*, pp. 195–213. McFarland, Jefferson, NC.

Digby, A. (2004) *Diversity and Division in Medicine: Health Care in South Africa from the 1800s*.

Evans-Pritchard, E. C. (1976) *Witchcraft, Oracles and Magic among the Zande*. Clarendon Press, Oxford.

Hilgard, E. R (1965) *Hypnotic Susceptibility*. Harcourt Brace, New York.

Humphrey, N. (1995) *Soul Searching*. Chatto & Windus, London.

Kaminer, W. (1999) *Sleeping with Extra-Terrestrials: The Rise of Irrationality and Perils of Piety*. Pantheon, New York.

Katz, S. (1978) *Mysticism and Philosophical Analysis*. Sheldon Press, London.

Lawrence, E. and Peters, E. (2004) Reasoning in believers in the paranormal. *Journal of Nervous and Mental Disorders*, 192, 1–7.

O'Keefe, D. (1982) *Stolen Lightning*. Robertson, Oxford.

Shermer, M. (1997) *Why People Believe Weird Things: Pseudoscience, Superstition and Other Confusions of Our Time*. Freeman, New York.

Subbotsky, E (1997) Explanations of unusual events: phenomenalistic causal judgments in children and adults. *British Journal of Developmental Psychology*, 15, 13–36.

Thomas, K. (1992) *Religion and the Decline of Magic*. Penguin, London.

de Waal Valfijt, A. (1968) Homo monstrous. *Scientific American*, October, 113–22.

Wheen, F. (2004) *How Mumbo-jumbo Conquered the World: A Short*

History of Modern Delusions. Fourth Estate, London.

Wiseman, R. (1996) Towards a psychology of deception. *Psychologist*, February, 61–4.

Wiseman, R. et al. (1995) Eyewitness testimony and the paranormal. *Skeptical Inquirer*, November/December, 29–32.

Wiseman, R., Greening, E. and Smith, M. (2003). Belief in the paranormal and suggestion in the seance room. *British Journal of Psychology*, 94, 285–97.

Wiseman, R., Watt, C., Stevens, P., Greening, E. and O'Keeffe, C. (2003) An investigation into alleged 'hauntings'. *British Journal of Psychology*, 94, 195–211.

TEN **Health**

Benson, H. (1996) *Timeless Healing: The Power and Biology of Belief*. Simon & Schuster, New York.

Gillick, M. R (1985) Common-sense models of health and disease. *New England Journal of Medicine*, 313, 700–3.

Evans, D. (2002) *Placebo: The Belief Effect*. Harper Collins, London.

Helman, C. G. (2000) *Culture, Health and Illness* (4th edition). Butterworth Heinemann, Oxford.

Kleinman, A. (1991) *Rethinking Psychiatry: From Cultural Category to Personal Experience*. Free Press, New York.

Landrine, H. and Klonoff, E. A. (1994) Cultural diversity in causal attributions for illness: the role of the supernatural. *Journal of Behavioural Medicine*, 17, 181–93.

Porter, R. (1997) *The Greatest Benefit to Mankind*. Harper Collins, London.

ELEVEN **Moral**

Blumenthal, W. M. (1998) *The Invisible Wall*. Counterpoint, Washington.

Browning, C. R. (1998) *Ordinary Men*. Penguin, London.

Glover, J. (2000) *Humanity: A Moral History of the Twentieth Century*. Yale University Press, New Haven, CT.

Suicide, terrorism, martyrdom and murder (2000) *Economist*, January 10, 18.

TWELVE **Science**

Berry, R. J. (1986) What to believe about miracles. *Nature*, 322, 321–2.

Berry, R. J. (1996) The virgin birth of Christ. *Science and Christian Belief*, 8, 101–10.

Dawkins, R. (1998) *Unweaving the Rainbow*. Allen Lane, London.

Gross, P. and Levitt, N. (1994) *Higher Superstitions: the Academic Left and its Quarrels with Science*. Johns Hopkins University Press, Baltimore.

Larson, E. J. and Witham, L. (1999) Scientists and religion in America. *Scientific American*, September, 78.

Lloyd, G. and Sivin, N. (2002) *The Way of the Word: Science and Medicine in Early China and Greece*. Yale University Press, New Haven, CT.

Nanda, M. (2003) *Prophets Facing Backwards: Postmodern Critiques of Science and Hindu Nationalism in India*. Rutgers, New Brunswick.

Park, R. (2000) *Voodoo Science*. Oxford University Press, Oxford.

Stannard, R. (1996) *Science and Wonders: Conversations about Science and Belief*. Faber and Faber, London.

Wolpert, L. (1993) *The Unnatural Nature of Science*. Faber and Faber, London.

Index

231